# Cytokine Therapies

## Novel Approaches for Clinical Indications

ANNALS OF THE NEW YORK ACADEMY OF SCIENCES

Volume 1182

# Cytokine Therapies

# Novel Approaches for Clinical Indications

*Edited by*
**RAYMOND P. DONNELLY**

*Published by Blackwell Publishing on behalf of the New York Academy of Sciences*
*Boston, Massachusetts*
*2009*

The *Annals of the New York Academy of Sciences* (ISSN: 0077-8923 [print]; ISSN: 1749-6632 [online]) is published 32 times a year on behalf of the New York Academy of Sciences by Wiley Subscription Services, Inc., a Wiley Company, 111 River Street, Hoboken, NJ 07030-5774.

**MAILING:** The *Annals* is mailed standard rate. POSTMASTER: Send all address changes to *ANNALS OF THE NEW YORK ACADEMY OF SCIENCES*, Journal Customer Services, John Wiley & Sons Inc., 350 Main Street, Malden, MA 02148-5020.

**Journal Customer Services:** For ordering information, claims, and any inquiry concerning your subscription, please go to interscience.wiley.com/support or contact your nearest office:

**Americas:** Email: cs-journals@wiley.com; Tel: +1 781 388 8598 or 1 800 835 6770 (Toll free in the USA & Canada).
**Europe, Middle East and Asia:** Email: cs-journals@wiley.com; Tel: +44 (0) 1865 778315
**Asia Pacific:** Email: cs-journals@wiley.com; Tel: +65 6511 8000
**Information for Subscribers:** The *Annals* is published in 32 issues per year. Subscription prices for 2009 are:
Print & Online: US$4862 (US), US$5296 (Rest of World), €3432 (Europe), £2702 (UK). Prices are exclusive of tax. Australian GST, Canadian GST and European VAT will be applied at the appropriate rates. For more information on current tax rates, please go to www3.interscience.wiley.com/aboutus/journal_ordering_and_payment.html#Tax. The price includes online access to the current and all online back files to January 1, 1997, where available. For other pricing options, including access information and terms and conditions, please visit www.interscience.wiley.com/journal-info.

**Delivery Terms and Legal Title:** Prices include delivery of print publications to the recipient's address. Delivery terms are Delivered Duty Unpaid (DDU); the recipient is responsible for paying any import duty or taxes. Legal title passes to the customer on despatch by our distributors.

**Membership information:** Members may order copies of *Annals* volumes directly from the Academy by visiting www.nyas.org/annals, emailing membership@nyas.org, faxing +1 212 298 3650, or calling 1 800 843 6927 (toll free in the USA), or +1 212 298 8640. For more information on becoming a member of the New York Academy of Sciences, please visit www.nyas.org/membership. Claims and inquiries on member orders should be directed to the Academy at email: membership@nyas.org or Tel: 1 800 843 6927 (toll free in the USA) or +1 212 298 8640.

Printed in the USA.

The *Annals* is available to subscribers online at Wiley InterScience and the New York Academy of Sciences' Web site. Visit www.interscience.wiley.com to search the articles and register for table of contents e-mail alerts.

ISSN: 0077-8923 (print); 1749-6632 (online)
ISBN-10: 1-57331-783-7; ISBN-13: 978-1-57331-783-2

A catalogue record for this title is available from the British Library.

ANNALS OF THE NEW YORK ACADEMY OF SCIENCES

*Volume 1182*

# Cytokine Therapies

## Novel Approaches for Clinical Indications

*Editor*

RAYMOND P. DONNELLY

This volume presents manuscripts stemming from the conference "Cytokine Therapies: Novel Approaches for Clinical Indications," held at the New York Academy of Sciences Conference Center on March 26–27, 2009, and cosponsored by the U.S. Food and Drug Administration. Sponsors were BioLegend, Invitrogen, and PBL InterferonSource. This activity was supported by educational donations provided by Abbott Laboratories, Amgen, Inc.; Celgene Corporation; and Centocor Research and Development, Inc.

## CONTENTS

# An Overview of Cytokines and Cytokine Antagonists as Therapeutic Agents

Raymond P. Donnelly,[a] Howard A. Young,[b] and Amy S. Rosenberg[a]

[a]*Division of Therapeutic Proteins, Center for Drug Evaluation and Research, Food and Drug Administration, Bethesda, Maryland, USA*

[b]*Laboratory of Experimental Immunology, Center for Cancer Research, National Cancer Institute, Frederick, Maryland, USA*

Cytokine-based therapies have the potential to provide novel treatments for cancer, autoimmune diseases, and many types of infectious disease. However, to date, the full clinical potential of cytokines as drugs has been limited by a number of factors. To discuss these limitations and explore ways to overcome them, the FDA partnered with the New York Academy of Sciences in March 2009 to host a two-day forum to discuss more effective ways to harness the clinical potential of cytokines and cytokine antagonists as therapeutic agents. The first day was focused primarily on the use of recombinant cytokines as therapeutic agents for treatment of human diseases. The second day focused largely on the use of cytokine antagonists as therapeutic agents for treatment of human diseases. This issue of the *Annals* includes more than a dozen papers that summarize much of the information that was presented during this very informative two-day conference.

*Key words*: cytokines; inflammation; interferons; interleukins; receptors

## Introduction

Cytokine therapies have great potential for treating a variety of diseases. These intercellular messengers activate numerous signaling pathways in virtually all cell types, but they are perhaps best known for their role in recruiting and activating various types of leukocytes in response to injury or infection. The actions of cytokines can be either positive or negative as concerns clinical context. For example, proinflammatory cytokines such as interleukin-1 (IL-1) and tumor necrosis factor-α (TNF-α) often contribute to the pathogenesis of autoimmune diseases, such as rheumatoid arthritis (RA) and inflammatory bowel disease (IBD), but appear to be essential for salutary responses to infectious agents. Similarly, cytokines can also enhance or inhibit the development of many cancers via their effects on pathways promoting tumorigenesis and tumor metastasis.

The therapeutic potential of cytokines was recognized decades ago, but attempts to use recombinant cytokines as conventional drugs have been mixed. Interleukin-10 (IL-10) is one example of a cytokine in which preclinical studies suggested great potential,[1] but therapeutic effectiveness was not demonstrated when tested in controlled human clinical trials.[2] After the clinical trials conducted in the 1990s failed to demonstrate significant improvement in patients with IBD, this cytokine was largely dropped from further clinical development. The unsuccessful use of cytokines, such as IL-10 and IL-12, as therapeutic agents was likely assured by using them like conventional drugs in terms of dosage and route of administration.

Address for correspondence: Raymond P. Donnelly, Ph.D., FDA CDER, Bldg 29A, Room 3B-15, 29 Lincoln Drive, Bethesda, MD 20892. Voice: 301-827-1776; fax: 301- 480-3256. Raymond.Donnelly@fda.hhs.gov

Cytokine Therapies: Ann. N.Y. Acad. Sci. 1182: 1–13 (2009).
doi: 10.1111/j.1749-6632.2009.05382.x © 2009 New York Academy of Sciences.

Because cytokines have activity on many different cell types and tissues, the serious adverse events associated with IL-12 therapy and the lack of efficacy associated with both IL-12 and IL-10 may well be attributable to the systemic rather than more localized administration of these cytokines.

Many of the early clinical trials of cytokines conducted in the 1990s, including the IL-10 and IL-12 trials, were conducted without a full understanding of cytokine function and regulation. As endogenous proteins, cytokines are regulated by various homeostatic mechanisms that can modulate their activities as therapeutic agents. They are also subject to biological degradation, and thus most have very short half-lives *in vivo*.

As discussed during this conference, scientists are now developing more effective methods to administer cytokines and to combine them with other treatments to maximize their effectiveness while minimizing their toxicities. They have developed novel ways to prolong the biological activity of these molecules and to deliver them in a more targeted manner to enhance efficacy and minimize toxicity. Furthermore, many new cytokines have been discovered since the time when the clinical trials of recombinant human IL-10 and IL-12 were conducted. Consequently, the number of defined cytokines has increased significantly as a result of advances in genomic database-mining techniques. For example, there are now 35 interleukins as well as several novel interferons (IFNs). Many of these more recently identified cytokines will be tested clinically in the near future, so the important lessons to be learned from the earlier clinical experiences with such cytokines as IL-2, IL-10, and IL-12 must be brought to bear with respect to development of these novel cytokines.

Many cytokines such as IL-1 and TNF-α are inherently toxic because of their marked proinflammatory activities. Thus these cytokines have proved excellent targets for development of antagonists that block their production or

activity. Consequently, many highly effective cytokine inhibitors have been developed during the last decade, including some very effective and highly selective inhibitors of IL-1 and TNF-α. It is possible, and perhaps likely, that many of the more recently discovered cytokines, such as IL-22 and IL-23, will also prove useful therapeutic targets for development of specific antagonists.

Gaining a more complete understanding of cytokine networks and developing more selective delivery methods are only two parts of the current challenge regarding these molecules. A third part is to convince academic, pharmaceutical, and regulatory scientists to agree on a common path forward that will maximize the possibility of clinical success of novel cytokine therapies. For example, regulatory requirements and intellectual property issues should not impede development and clinical testing of potentially life-saving therapies for patients with unmet medical needs. The FDA's Critical Path Initiative (http://www.fda.gov/ScienceResearch/SpecialTopics/CriticalPathInitiative) provides a useful platform to help bring these distinct stakeholders together to discuss how to rejuvenate efforts to test promising cytokines, as well as to shorten the lag between initial drug discovery and regulatory approval.

To discuss current topics regarding the clinical use of cytokines and cytokine antagonists, scientists and clinicians from academia, the biopharmaceutical industry, and FDA met together at the New York Academy of Sciences on March 26–27, 2009. The goal of this two-day conference was to provide a forum to review lessons learned from previous clinical trials of many cytokines and cytokine antagonists and to consider novel approaches to improve the effectiveness of cytokines as therapeutic agents in future clinical trials. The program was co-organized by Drs. Raymond Donnelly and Amy Rosenberg of the FDA (Bethesda, MD) and Howard Young of the National Cancer Institute (Frederick, MD).

## Day 1: Cytokines as Therapeutic Agents

### Interferons as Agents for Treating Infectious Diseases

Cytokines play an important role in the host response to microbial pathogens, such as bacteria and viruses. The interaction of microbial pathogens with cell surface receptors, such as the *Toll*-like receptors (TLRs) or the entry of viruses into a cell often activates transcription of many cytokine genes and production of the corresponding proteins. These cytokines are often secreted by infected cells and deliver secondary warning signals to nearby cells to activate their innate antimicrobial defenses. Cytokines can also act on many of the cell types that constitute the host immune system, including B cells, T cells, natural killer (NK) cells, and macrophages. One such antiviral cellular response occurs when the cytokine IFN-α binds to its receptors. IFN-α receptors are expressed

on virtually all somatic cell types, and binding of IFN-α to its cognate receptor complex induces expression of many so-called interferon-stimulated genes (ISGs). Many of the protein products encoded by these genes then act to induce antiviral activity in surrounding cells.

IFNs are a group of cytokines that can induce potent antiviral activity in many cell types. Type-I IFNs include IFN-α, IFN-β, and IFN-ω, all of which bind and signal through a common IFN-α receptor complex. There are thirteen types of human IFN-α, one type of IFN-β, and one type of IFN-ω. There is only one type-II IFN, known as IFN-γ, which binds and signals via the distinct IFN-γ receptor complex (Fig. 1). More recently, a novel group of IFNs was discovered that makes up a third family of IFNs now known as the type-III IFNs.[3,4] This new group of IFNs includes three members: IFN-λ1, -λ2, and -λ3. These cytokines bind to a distinct receptor denoted IFN-λR1 (also known as IL-28RA).[3,4]

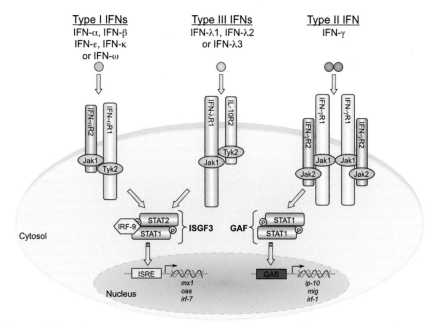

**Figure 1.** Interferons are a group of cytokines that can induce potent antiviral activity in many cell types. The type I, type II, and type III interferons bind to distinct receptor complexes on the cell membrane. Signal transduction induced by the binding of interferons to their cognate receptors induces expression of interferon-stimulated genes (ISGs). The proteins encoded by these genes in turn mediate the antiviral activity of the interferons.

IFN-α is an approved treatment for chronic hepatitis C virus (HCV) infection. If left untreated, HCV infection can progress to liver cirrhosis and eventually cancer (hepatoma) in a subset of patients. The current standard of care for the treatment of chronic HCV infection is to administer a pegylated form of IFN-α in combination with the antiviral drug ribavirin. Addition of the polyethylene glycol (PEG) moiety to IFN-α prolongs the half-life of the protein in the body. Unfortunately, this treatment regimen (pegylated IFN-α plus ribavirin) significantly reduces viral load in only about 50% of patients who are infected with HCV genotype-1. For additional information, please see: http://digestive.niddk.nih.gov/ddiseases/pubs/chronichepc/index.htm.

One approach to improving the efficacy of hepatitis C treatment regimens is to increase the overall bioactivity of IFN-α. Each of the naturally occurring IFN-α molecules has a slightly different efficacy profile, suggesting that a chimeric protein might be even better than the native isoforms. Currently, there is one FDA-approved IFN-α product, known as Infergen, that was derived from a consensus sequence of the 12 naturally occurring human IFN-α proteins. There are also two versions of recombinant human IFN- α2, known as Roferon and Intron A, available for clinical use in the United States.

Dr. Julian Symons of Roche-Palo Alto, LLC, and his colleagues investigated whether a novel technology known as "gene shuffling" could be used to increase the ability of IFN-α to stimulate kill of viruses and modulate the immune response.[5] Libraries of gene-shuffled IFN molecules were tested for their antiviral, immunomodulatory, and antiproliferative activities. After several rounds of selection, they chose the candidate that had the strongest antiviral activity. This designer IFN protein demonstrated enhanced antiviral activity *in vitro* when compared to other IFN-α proteins. Consequently, this hybrid IFN was tested in a small clinical trial in patients with chronic HCV infection. Unfortunately, this protein induced

production of anti–IFN-α antibodies in many patients, and these antibodies neutralized the biological activity of the drug. Further development of this cytokine as a novel drug candidate was halted. The development of anti–IFN antibodies in patients who were treated with this hybrid IFN molecule may have been caused by the fact that this protein contained novel, non-native stretches of amino acids that were perceived as foreign by the host immune system. Such responses may extend to conserved regions of the molecule and produce neutralizing responses in a process called *epitope spreading*. This clinical experience underscores the profound impact that even a few amino acid differences can impart on the relative immunogenicity of a protein.

Conventional antiviral therapy with IFN-α products, such as Pegasys (peginterferon α-2a) or PegIntron (peginterferon α-2b) often induces many undesirable side effects, including headache, fevers, chills, nausea, and myelosuppression. Consequently, scientists are seeking alternative, less toxic agents to treat this disease. One promising, novel, antiviral agent that was discussed by Dr. Dennis Miller of ZymoGenetics, Inc. (Seattle, WA) is IFN-λ. IFN-λ, also known as IL-29, is functionally similar to IFN-α but is potentially much less toxic than IFN-α as a treatment for chronic HCV infection.[6]

Both IFN-α and IFN-λ are induced by viruses *in vitro* and *in vivo*, but they bind to distinct receptors (Fig. 1). The IFN-λ receptors are expressed on fewer cell types than are those for IFN-α, However and critically, like IFN-α receptors, IFN-λ receptors are expressed on liver cells, the primary cellular targets of the HCV. The fact that expression of IFN-λ receptors is much more restricted than IFN-α receptors suggests the possibility that IFN-λ will induce fewer side effects than IFN-α. In terms of receptor binding kinetics, IFN-α binds and releases from its receptor relatively quickly, whereas IFN-λ binds and releases from its receptor more slowly. Both cytokines induce expression of the same subset of ISGs and appear to share common antiviral, antiproliferative, and apoptosis-inducing activities.[7,8]

In a Phase-1 clinical trial, ZymoGenetics's pegylated version of IFN-λ1 (IL-29) was well tolerated and did not induce the neutropenia, thrombocytopenia, or anemia that commonly occur in patients who are treated with IFN-α products. Several patients in their initial Phase-1 clinical trial demonstrated significant decreases in HCV RNA levels, and the sponsor plans to continue studying this novel drug in a Phase-2 clinical trial.[6]

Cytokines, particularly the IFNs, have the potential to treat many other infectious diseases in addition to hepatitis C. For example, as discussed by Dr. Howard Young and his colleagues at the National Cancer Institute (Frederick, MD), IFN-γ has been shown to be a very effective agent for treating several types of infectious disease, including drug-resistant tuberculosis.[9] However, this cytokine activates many inflammatory responses throughout the body and induced an unacceptable level of adverse effects in these trials. Dr. Steven Holland of the National Institute of Allergy and Infectious Diseases (NIAID) pointed out that, after many unsuccessful clinical trials for other indications, because of toxicity, IFN-γ was eventually approved by the FDA only as a treatment for two, rare genetically inherited conditions: chronic granulomatous disease (CGD) and osteopetrosis.

Despite these limited clinical applications, continued exploration of the role of endogenous IFN-γ in the physiologic response to infectious agents has led to new treatment options for people with specific inherited defects in the IFN-γ receptor gene.[10] For example, a very small number of people do not express IFN-γ receptors and as a result are unable to mount an effective immune response against certain microbial pathogens, including some types of mycobacteria. Holland and his team successfully treated one such patient suffering from disseminated *Mycobacterium avium* by treating her with IFN-α.[11] However, they also found that IFN-α cannot substitute for IFN-γ in all types of infections. A greater understanding of cytokine-induced signaling pathways might help clinicians devise more specific treatments for various diseases.

## Interleukin-7 as an Immune System Rejuvenator

The study of cytokine mechanisms can reveal not only treatments for specific diseases but also ones with broad potential to rejuvenate or reinforce the immune system. "One cytokine with broad immunostimulatory potential is IL-7," said Dr. Crystal Mackall of the Pediatric Oncology Branch of the National Cancer Institute. IL-7 is a member of a broader subset of cytokines that also includes IL-2, IL-4, IL-9, IL-15, and IL-21. Each of these cytokines binds initially to its unique ligand-binding (alpha) chain. However, they all share the use of the IL-2 receptor common gamma chain (γc) as a secondary receptor component that is essential for signaling.

According to Dr. Mackall, IL-7 can restore certain T-cell populations that are diminished naturally over time. Normally, these T cells are replenished in two ways, either via a thymic-dependent process or via a thymic-independent process called homeostatic peripheral expansion (HPE).[12]

IL-7 predominantly promotes the regeneration of T cells via the non-thymic-dependent process. Thus, IL-7 could potentially be of benefit to patients who have undergone thymic involution, including the elderly, since as people age, the thymus becomes less efficient as a source of T-cell generation. A non-thymus-dependent method of generating T cells might also increase the levels of such cells in patients in whom T cells have been depleted by treatments, such as those undergoing chemotherapy and bone marrow transplantation, or by infection such as those with HIV infection. IL-7 preferentially expands naive T cells and increases overall T-cell repertoire diversity. No serious adverse effects were observed during the initial human clinical trials of IL-7.[12]

## Cytokines as Anticancer Agents

### A Fresh Look at IL-2

The use of cytokines, such IL-2, as a treatment for certain types of cancer results in complete remission in a very small subset of patients and a significant but less than full response in a small number as well. It is clear that more research needs to be done in order to learn how to extend this limited clinical success to a larger number of patients.[13]

The antiproliferative properties of cytokines are mediated via a number of mechanisms. They can induce direct antiangiogenic effects or indirectly induce expression of genes that are antiproliferative. Cytokines, in general, and IL-2 in particular, can also induce apoptosis, a type of programmed cell death, either directly by activation-induced cell death, requiring expression of the death receptors Fas and Fas ligand, or indirectly via activation of cytotoxic T lymphocytes that kill cells via an apoptotic mechanism.

"In the early 1990s, IL-2 raised great hopes of providing a treatment or perhaps even a cure for some types of cancer," said Dr. Michael Lotze of the University of Pittsburgh Cancer Research Institute. Treatment with this cytokine induces a durable remission rate in a small subset of patients (8–10%) with melanoma and renal cell carcinoma.[13]

However, the small percentage of patients in which success is demonstrated is not sufficiently robust. Therefore, work was focused on defining the mechanism by which this therapy works in that small subset of responders. Also, given the severe toxicities associated with IL-2 treatment, studies have also focused on profiling patients that might truly benefit from the treatment. A recent study by Howard Kaufman's lab examined responsiveness and nonresponsiveness to IL-2 therapy and found that individuals with high levels of the growth factor, VEGF, prior to therapy were less likely to respond compared to patients with low concentrations of VEGF in their bloodstream.[14] Similarly, fibronectin was high in nonresponders but low in responders. These levels could perhaps be used to prospectively identify patients who will respond to IL-2.

Another way to understand the low rate of success with IL-2 therapy is to examine its activity in the context of the cell's response to certain growth factors and stress. "Cancer can be considered to be a metabolic disorder because the body's metabolism supports tumor formation," said Lotze. The immune system is complicit because it tolerates formation of the tumor and surrounding new blood vessels. Scientists have known for some time that apoptosis is disabled during tumor formation. Lotze believes that a competing cell death process called *autophagy*, in which cells catabolize their own cellular components when cells are under survival stress, allows many tumor cells to survive anticancer treatments. He further noted that VEGF and HMGB1 are important switches that modulate apoptosis and autophagy.[13] This would suggest that the high levels of VEGF associated with poor response to IL-2 may interfere with apoptosis by favoring autophagy and would further suggest use of monoclonal antibodies (mAbs) to VEGF as a mechanism to facilitate beneficial effects of IL-2 in patients with high VEGF levels.

To improve the remission rate of IL-2, we need to acquire a better understanding of the spectrum of IL-2 effects on cellular and soluble mechanisms that affect survival and death of tumor cells. In addition to induction of apoptosis, IL-2 directly activates NK cells and macrophages, promotes Th1 cell activity, and induces proliferation of B cells. IL-2 therapy for cancer is also associated with induction of autoimmunity, which, in the setting of IFN-α treatment, appears to correlate with clinical benefit.

### Defining the Mechanisms by Which Cytokines Mediate Their Activity

Dr. Ahmad Tarhini of the University of Pittsburgh Cancer Institute described the clinical and immunological basis of IFN-α therapy in melanoma patients.[15] Melanoma is a highly curable cancer if diagnosed and treated aggressively in the early stages of development but is usually fatal if allowed to metastasize. Some spontaneous regressions occur, often in the setting of autoimmune phenomena, suggesting that the immune system can suppress this disease, perhaps through the involvement of T cells, macrophages, and NK cells.

Tarhini and his colleagues are attempting to identify biomarkers that may help to predict which patients will respond favorably to IFN-α treatment. They have found that patients with high pretreatment levels of certain proinflammatory cytokines, such as IL-1α, MIP-1α, IL-6, and TNF-α are more likely to have relapse-free survival. They also observed that survival of melanoma patients following IFN-α treatment was greater in patients who developed autoimmunity such as vitiligo or thyroiditis.[15] These data speak to autoimmunity or a high potential for its generation, as a means to effectively target and kill tumor cells. Indeed, the presence of T-cell infiltrates in the tumor is a positive prognostic marker, and the presence of T-cell infiltrates within regional nodal metastases often predicts positive responsiveness to IFN-α treatment.[16,17]

Dr. Ernie Borden of the Cleveland Clinic underscored the need to understand the intracellular mechanisms that occur downstream of IFN binding to its cell-surface receptor. Borden indicated that, to fully exploit the therapeutic potential of IFNs, we must gain a better understanding of the regulation and function of the more than 300 ISGs that are induced by this cytokine. Some of these genes, such as TRAIL and XAF1, are proapoptotic, whereas other genes such as G1P3 (ISG 6–16) inhibit apoptosis. Many other IFN-inducible genes are immunomodulatory or antiangiogenic. Defining the precise function(s) of ISGs may help scientists to overcome the resistance mechanisms and drug-related toxicities that are associated with IFN-α therapy. Such studies may also reveal improved ways to enhance the antitumor activity of the IFNs.

One common downstream signaling pathway that is activated by IFN-α is the JAK/STAT pathway. A drug known as stibogluconate (SSG) enhances activation of the JAK/STAT pathway by IFN-α and thereby increases the magnitude of activity induced by IFN-α. The combination of stibogluconate and IFN-α is now being tested in a Phase 1 clinical trial in melanoma patients.[18]

## IL-21 in Addition to Monoclonal Antibodies for the Treatment of Cancer

IL-21, a relative newcomer in the field of cytokines, is also being tested as an anticancer therapeutic agent.[19] This cytokine is secreted by activated CD4+ T cells and NK cells. It helps regulate immunoglobulin production and Ig-isotype switching by B cells and has activating effects on macrophages. IL-21 has demonstrated anticancer properties in tumor-modeling studies in mice as well as in early clinical trials in humans. In his presentation, Dr. William Carson (Ohio State University) pointed out that IL-21 has structural homology to other class-I cytokines, including IL-2. His group has shown that this cytokine can induce antitumor responses in murine models of melanoma, renal cell carcinoma, colon adenocarcinoma, breast cancer, and other tumors. The antitumor effects of IL-21 appear to be mediated largely by NK cells and CD8+ T cells.[19]

Dr. Carson and his colleagues evaluated the antitumor activity of recombinant IL-21 in combination with trastuzumab (Herceptin, Genentech, Inc.), a monoclonal antibody that inhibits the growth of Her2/Neu-positive tumors and mediates antibody-dependent cellular cytotoxicity (ADCC). They found that IL-21 enhances NK cell–mediated ADCC and cytokine production when administered in combination with several different monoclonal antibodies, such as rituximab (Rituxan, Genentech) or trastuzumab. These findings suggest that combined treatment with the cytokine IL-21 plus an appropriate monoclonal antibody may yield more robust and sustained antitumor responses than can be achieved by monotherapy with either agent alone.[19]

### Lessons Learned

Since the time when many of the early, cytokine-based clinical trials were first conducted (1990–2000), numerous technological advances have occurred that can revive the clinical potential of at least some of these cytokines as drug candidates. These include advances in drug delivery platforms and methods to prolong the *in vivo* half-life of therapeutic proteins. Additionally, biomarkers can now be used to prospectively identify patients who would be more likely to respond to a particular cytokine. Many of the new analytical tools that exist today, particularly microarray technologies, can also be used to identify better biomarkers of cytokine-mediated pharmacodynamic activities.

Some of the cytokines that failed in clinical trials in the 1990s might now be worth reevaluating. A good example of one such cytokine is IL-12, which was largely abandoned after the original clinical studies of this cytokine were conducted in the 1990s. This cytokine was evaluated as a potential treatment for certain types of cancer and infectious disease but was found to be highly toxic and largely ineffective as a mono-therapeutic agent in several clinical trials. However, it is now clear that IL-12 might be more effective as an anticancer agent if administered at lower, less toxic concentrations together with other anticancer drugs or cytokines.

### Day 2: Cytokines as Therapeutic Targets

Inflammatory autoimmune diseases such as RA and psoriasis can often be treated successfully with cytokine antagonists, such as the TNF inhibitors, Enbrel, or Remicaide. These biological agents may also be useful for treating other clinical indications for which there are currently no alternative treatment options. In addition, many novel cytokines have been discovered during the last ten years, and it is reasonable to predict that at least some of these cytokines will provide useful therapeutic targets for development of novel cytokine antagonists.[20]

### Cytokines and Cytokine Antagonists for the Treatment of Autoimmune and Inflammatory Diseases

### Key discussion points:

- Proinflammatory cytokines induce many of the pathogenic processes that are characteristically associated with many autoimmune diseases.
- Monoclonal antibodies that block the activity of specific cytokines, either by blocking cytokine receptors or neutralizing cytokine activity, are often highly effective for the treatment of such autoimmune diseases as RA, IBD, and psoriasis.
- Gene expression profiling may help scientists to better understand how cytokines regulate the inflammatory processes that are associated with many autoimmune diseases.
- New methods for targeting and delivering cytokines and cytokine antagonists have the potential to increase the therapeutic utility of these agents and reduce their undesirable toxicities.

Cytokines often function as intercellular messengers to activate the immune system, and they play a central role in many diseases that involve the immune system. These diseases include inflammatory diseases characterized by excessive production of inflammatory cytokines, such as TNF-$\alpha$ and/or IL-1, and autoimmune diseases that are characterized by immune responses directed against the body's own proteins or cells. Effective treatment of many immune-mediated disorders has been improved greatly by the discovery of cytokine inhibitors. These include cytokine receptor constructs that can bind and neutralize specific cytokines and monoclonal antibodies that target specific cytokines.

RA is a classic example of an inflammatory disease where cytokines play a prominent role. RA is characterized in part by the infiltration and activation of inflammatory T cells that produce proinflammatory cytokines, such as TNF-α, IL-1, and IL-6. These cytokines, in turn, mediate the activation of tissue-destroying metalloproteinases and expression of vascular adhesion molecules that recruit lymphocytes, macrophages, and other types of leukocytes to the joints. As a consequence of these events, B cells often become activated and produce autoantibodies. If not blocked pharmaceutically, these processes can lead to progressive joint destruction.

## Rheumatoid Arthritis

For many patients, RA can also affect organ systems other than the joints. For example, patients can also develop cardiovascular disease, chronic pulmonary obstructive disease (COPD), blood disorders, neurological symptoms, pulmonary effects, and ocular problems. Dr. Larry Moreland of the University of Pittsburgh reviewed ways to specifically target TNF-α, a proinflammatory cytokine that plays a central role in the pathogenesis of RA.[21]

TNF-α appears to play a central role in disease activity in roughly three-quarters of all RA patients. Currently, five biologic agents that inhibit TNF-α are approved for clinical use in the United States. Three are mAbs against TNF-α, and two are soluble receptor constructs that act by binding TNF-α and facilitating its clearance from the body. The TNF inhibitors such as Enbrel and Remicaide have been shown to decrease symptoms, slow disease progression, and improve the quality of life for many patients with RA.[21]

It is unclear how frequently patients develop neutralizing antibodies to TNF antagonists such as Enbrel or Remicaide after being treated with one or more of these drugs for some period of time. Comparative studies to accurately quantify the incidence of anti-TNF antibodies in RA patients treated with different anti-TNF agents have not yet been reported. Furthermore, TNF-α may not be the most appropriate cytokine to target in all RA patients. Other cytokines, such as IL-1, IL-6, or IL-12 may be the dominant disease-driving cytokine in RA patients who do not respond well to TNF inhibitors. The use of contemporary cytokine-profiling platforms could provide very useful information regarding which patients will respond most favorably to specific anticytokine therapies.

## Blocking IL-12 and IL-23 as a Therapy for Autoimmune Diseases

Psoriasis is an inflammatory disease that is characterized in part by the rapid growth of skin cells. Scientists have found that in classic psoriasis T cells often become activated to produce multiple cytokines that stimulate the growth of keratinocytes. One such cytokine is IL-12, which acts on Th1 cells to induce production of IFN-γ. Another highly related cytokine, IL-23, acts preferentially on Th17 cells to stimulate production of a distinct subset of proinflammatory cytokines, including IL-6, IL-17, and IL-22.

Michael Elliott of Centocor, Inc. discussed clinical development of ustekinumab (Stelara), a humanized mAb that binds the shared p40 subunit of IL-12 and IL-23.[22] This mAb binds with high affinity to the p40 subunit of IL-12 and IL-23 and prevents these cytokines from binding to the IL-12Rβ1 receptor. This in turn prevents activation of the intracellular signaling cascade that is normally activated by these cytokines. This novel cytokine antagonist is approved for marketing in Canada and Europe for moderate to severe plaque psoriasis. It has also recently been approved by the FDA for use in the United States.

By blocking both IL-12 and IL-23, ustekinumab inhibits inflammatory cell infiltration, decreases expression of such cytokines as IFN-γ, IL-17, and IL-22, and reduces epidermal hyperplasia. This drug does not appear to affect the levels of circulating Th1, Th2, Treg, or NK cells, although there may be some reduction in the levels of circulating Th17 cells. In addition to its use as a treatment for chronic psoriasis, ustekinumab may also be useful as a treatment for other autoimmune diseases such as psoriatic arthritis and Crohn's disease.[22]

## Interferon-β in Autoimmune Disease

Richard Ransohoff of the Cleveland Clinic Foundation pointed out that recombinant human IFN-β is very effective as a treatment for multiple sclerosis (MS) in at least a subset of patients.[23] However, treatment with recombinant human IFN-β is not uniformly effective in all MS patients, and it has several negative features. It is expensive, inconvenient, and induces many undesireable side effects. Nevertheless, prior to its approval by the FDA for clinical use as a treatment for MS, there were very few treatments options for patients with this disease.

Scientists would like to be able to predict which patients will respond positively to IFN-β therapy. To identify predictive biomarkers, Ransohoff and his colleagues evaluated gene expression profiles from multiple MS patients to see if individuals treated with IFN-β had varied responses to this cytokine. They used microarrays containing a large subset of ISGs to examine changes in gene expression levels following IFN-β treatment.[23]

They found that neither the magnitude nor the stability of the biological response to IFN-β was responsible for differences in responsiveness among patients. They concluded that a specific gene or group of genes that are induced by IFN-β must account for the differential responsiveness of patients to IFN-β therapy. Their group and others have recently identified a number of genes that are commonly associated with MS, and they are currently exploring how these genes are regulated by IFN-β treatment.[23]

Another team that is using gene expression profiling to learn more about how cytokines influence diseases is headed by Dr. Virginia Pascual of the Baylor Institute for Immunology Research. Pascual and her team are looking for gene signatures associated with systemic lupus erythematosus (SLE). They have determined that ISG signatures provide useful biomarkers for diagnosis and assessment of disease activity in SLE.[24]

## IL-1 in Autoinflammatory Disease

Research in rare genetic diseases can help expand our understanding of inflammatory processes in more common diseases, explained Dr. Raphaela Goldbach-Mansky of the Translational Autoinflammatory Disease Section at the National Institute of Arthritis and Musculoskeletal and Skin Diseases (NIAMS).[25] Three such rare diseases are collectively called cryopyrin-associated periodic syndromes (CAPS). Familial cold autoinflammatory syndrome (FCAS) involves cold-induced attacks of fever, neutrophilic urticaria, conjunctivitis, and joint pain, lasting 12 to 24 hours and then resolving. Muckle Wells Syndrome (MWS) is a more severe and persistent disease that is not cold induced and involves fever, neutrophilic urticaria, joint pain, progressive hearing loss, and amyloidosis. The third disorder is neonatal onset multisystem inflammatory disease (NOMID), involving the same symptoms as MWS, but including bony overgrowth of the knees, organ damage, and mental retardation.

All three of these diseases appear to be mediated by the proinflammatory cytokine, IL-1. A recombinant human IL-1 receptor antagonist, (anakinra, Kineret™), first approved by the FDA in 2001 for the treatment of RA, provides a proven treatment for these diseases. The effectiveness of blocking IL-1 in NOMID was demonstrated by the findings that treatment with anakinra in patients with NOMID resulted in immediate resolution of the skin rash, and the symptoms returned when the IL-1 inhibitor was withdrawn. In patients with NO-MID, blocking IL-1 helped restore hearing and vision in some patients but had no effect in others. However, IL-1 inhibitors did not help prevent the growth of bony lesions in the knees. Early diagnosis and active treatment with an IL-1 inhibitor, such as anakinra, can reduce and perhaps prevent the development of organ-specific damage and disability.[25]

IL-1β plays a central role in several inflammatory diseases, including RA, CAPS, and gout, said Neil Stahl of Regeneron Pharmaceuticals, Inc. Rilonacept, a drug that is approved for treating CAPS and is now in Phase-3 trials for gout, is a receptor-Fc fusion protein that traps and promotes clearance of IL-1.[26] It is very specific for IL-1 and has a very high affinity for this cytokine. Rilonacept and IL-1β form a complex that prevents the biological activity of this cytokine. Stahl and his colleagues measured the levels of circulating IL-1 in several diseases and found that IL-1β levels are highest in patients with CAPS (FCAS), followed by gout and then RA. Normal healthy volunteers appear to have very low IL-1β synthesis.

The Regeneron group concluded that studying IL-1β:Rilonacept complex levels may prove useful in identifying IL-1β "driven" diseases and therefore diseases that are more responsive to IL-1 inhibition. Other scientists at Regeneron, led by Allen Radin, are exploring the use of Rilonacept as a treatment for chronic gouty arthritis, a rare subset of the gout spectrum that is resistant to the standard drugs that are used to treat gout.[26]

## Use of Bioengineering to Improve the Bioactivity and Tissue-Specific Targeting of Cytokines as Drugs

Although many cytokines have defined biological activities that can be harnessed to treat certain diseases, there are at least two major challenges regarding the clinical use of cytokines as therapeutic agents. First, most cytokines have short half-lives (typically a few hours at most) when injected *in vivo*, and second, they often induce many undesired side effects, because receptors for most cytokines are broadly expressed on many different cells/tissues throughout the body. In their native form, cytokines are usually eliminated rapidly, both by receptor-mediated uptake as well as by enzymatic inactivation by proteases. The short half-lives of these biologic agents significantly limit their efficacy. Increasing the stability of these proteins would allow these agents to be given less frequently and at lower doses. Consequently, scientists have developed several methods to extend the half-life of cytokines *in vivo* and target them more selectively to specific tissue/organs.

## Polyethylene Glycol Extends the Half-Life of Cytokines

The addition of polyethylene glycol (PEG) to proteins can greatly increase their half-life in the body and enhance other pharmacologically important properties such as their solubility. "Pegylation" was originally pioneered by Abraham Abuchowski of Prolong Pharmaceuticals, Inc., while he was a Ph.D. student at Rutgers University.[27] It is now a standard and highly accepted method in the biopharmaceutical industry that is widely used to prolong the half-life of many protein therapeutics.

Pegylation has a number of benefits. It is nontoxic and increases the circulating half-life of many drugs, thereby reducing the number of doses and the frequency of dosing. Adding PEG causes molecules to become more water soluble because PEG readily binds water, creating a hydrodynamic shell that though it diminishes binding of a cytokine to its receptor does not fully block binding and with the increased longevity of the product produces a more sustained effect. Numerous chemical methods have now been developed to facilitate attachment of PEG moieties to both protein and nonproteinaceous molecules.

## Cytokine Delivery Using Food-Grade Bacteria

An innovative and novel delivery system that is now being evaluated clinically involves oral delivery of therapeutic peptides or proteins via genetically engineered bacteria. Pioneered by Dr. Lothar Steidler and his team at ActoGeniX NV in Belgium, these noninvasive, noncolonizing, food-grade bacteria can secrete bioactive proteins or peptides into the gastrointestinal tract. These ActoBiotics are produced by a type of bacteria called *Lactococcus lactis* that are engineered to express a specific cytokine(s).[28]

The company is using this novel delivery method to deliver recombinant human IL-10 to the gut as a possible new treatment of IBD.[29] This cytokine was tested in the past for treatment of IBD, but it was administered as a parenteral bolus either intravenously or subcutaneously. In view of its short half-life, the widespread expression of IL-10 receptors, and the systemic route of administration, the protein probably did not reach the critical target tissue, the mucosal lining of the gut. In the clinical trials of recombinant human IL-10 that were performed in the mid-1990s, parenteral administration of IL-10 at high concentrations proved to be ineffective. However, the use of *Lactococcus* to deliver IL-10 orally may now provide an improved method to deliver this cytokine more selectively to the gut where it may act to suppress inflammation.

## Final Comments

As discussed by Dr. Steven Kozlowski of the FDA Office of Biotechnology Products, a number of technical hurdles still limit the full clinical potential of cytokine-related therapies.[30] However, much has been learned during the last 20 years or so as a result of the initial attempts to use recombinant cytokines as therapeutic agents for various clinical indications.

Clearly, much more research needs to be done before we fully understand how cytokines can be most effectively used as therapeutic agents for treating the spectrum of human diseases. However, cytokines and cytokine inhibitors continue to provide a rich pipeline of novel therapeutic agents for treating various cancers, autoimmune diseases, and infectious diseases. On the basis of discussions that took place during this conference, there was general agreement that a greater understanding of the downstream, intracellular, signaling cascades and cytokine-inducible genes will facilitate development of more judicious clinical use of cytokines as therapeutic agents. Several of the speakers at this conference pointed out that many new biomarkers have now been discovered that may provide very useful tools to identify specific patient subpopulations that would be predicted to have therapeutic benefit from a particular cytokine or cytokine antagonist and to monitor therapeutic efficacy.

## Conflicts of interest

The authors declare no conflicts of interest.

## References

1. Kühn R, J. Löhler, D. Rennick, K. Rajewsky & W. Müller. 1993. Interleukin-10-deficient mice develop chronic enterocolitis. *Cell* **75:** 263–274.
2. Bickston, S.J. & F. Cominelli. 2000. Recombinant interleukin-10 for the treatment of active Crohn's disease: lessons in biologic therapy. *Gastroenterology* **119:** 1781–1783.
3. Sheppard, P. *et al.* 2003. IL-28, IL-29 and their class II cytokine receptor IL-28R. *Nat. Immunol.* **4:** 63–68.
4. Kotenko, S.V. *et al.* 2003. IFN-lambdas mediate antiviral protection through a distinct class II cytokine receptor complex. *Nat. Immunol.* **4:** 69–77.
5. Brideau-Andersen, A.D. *et al.* 2007. Directed evolution of gene-shuffled IFN-alpha molecules with activity profiles tailored for treatment of chronic viral diseases. *Proc. Natl. Acad. Sci. USA* **104:** 8269–8274.
6. Miller, D.M. *et al.* 2009. Interferon lambda (IFN-λ) as a potential new therapeutic for hepatitis C. *Ann. N. Y. Acad. Sci.* **1182:** 80–87.
7. Doyle, S.E. *et al.* 2006. Interleukin-29 uses a type-1 interferon-like program to promote antiviral

responses in human hepatocytes. *Hepatology* **44:** 896–906.

8. Marcello, T. *et al*. 2006. Interferons α and λ inhibit hepatitis C virus replication with distinct signal transduction and gene regulation kinetics. *Gastroenterology* **131:** 1887–1898.

9. Miller, C.H.T., S.G. Maher & H.A. Young. 2009. Clinical use of interferon-γ. *Ann. N. Y. Acad. Sci.* **1182:** 69–79.

10. Holland S.M. 2007. Interferon gamma, IL-12, IL-12R and STAT1 immunodeficiency diseases: disorders of the interface of innate and adaptive immunity. *Immunol. Res.* **38:** 342–346.

11. Ward, C.M. *et al*. 2007. Adjunctive treatment of disseminated *Mycobacterium avium* complex infection with interferon alpha-2b in a patient with complete interferon-gamma receptor R1 deficiency. *Eur. J. Pediatr.* **166:** 981–985.

12. Sportès, C., R.E. Gress & C.L. Mackall. 2009. Perspective on potential clinical applications of recombinant human interleukin-7. *Ann. N. Y. Acad. Sci.* **1182:** 28–38.

13. Chavez, A.R. *et al*. 2009. Pharmacologic administration of interleukin-2: inducing a systemic autophagic syndrome? *Ann. N. Y. Acad. Sci.* **1182:** 14–27.

14. Sabatino, M. *et al*. 2009. Serum vascular endothelial growth factor and fibronectin predict clinical response to high-dose interleukin-2 therapy. *J. Clin. Oncol.* **27:** 2645–2652.

15. Tarhini, A.A. & J.M. Kirkwood. 2009. Clinical and immunologic basis of interferon therapy in melanoma. *Ann. N. Y. Acad. Sci.* **1182** 47–57.

16. Franzke, A. *et al*. 1999. Autoimmunity resulting from cytokine treatment predicts long-term survival in patients with metastatic renal cell cancer. *J. Clin. Oncol.* **17:** 529–533.

17. Gogas, H. *et al*. 2006. Prognostic significance of autoimmunity during treatment of melanoma with interferon. *N. Engl. J. Med.* **354:** 709–718.

18. Fan, K., E. Borden & T. Yi. 2009. Interferon-gamma is induced in human peripheral blood immune cells *in vitro* by sodium stibogluconate/interleukin-2 and mediates its anti-tumor activity in vivo. *J. Interferon Cytokine Res.* **29:** 451–460.

19. Bhave, N.S. & W.E. Carson III. 2009. Immune modulation with interleukin-21. *Ann. N. Y. Acad. Sci.* **1182:** 39–46.

20. Scheinecker, C., K. Redlich & J.S. Smolen. 2008. Cytokines as therapeutic targets: advances and limitations. *Immunity* **28:** 440–444.

21. Moreland, L.W. 2009. Cytokines as targets for anti-inflammatory agents. *Ann. N. Y. Acad. Sci.* **1182:** 88–96.

22. Elliott, M. *et al*. 2009. Ustekinumab: lessons learned from targeting IL-12/23p40 in immune-mediated diseases. *Ann. N. Y. Acad. Sci.* **1182:** 97–110.

23. Rani, M.R.S. *et al*. 2009. Heterogeneous, longitudinally-stable molecular signatures in response to interferon-β. *Ann. N. Y. Acad. Sci.* **1182:** 58–68.

24. Chaussabel, D. *et al*. 2008. A modular analysis framework for blood genomics studies: application to systemic lupus erythematosus. *Immunity* **29:** 150–164.

25. Goldbach-Mansky, R. 2009. Blocking interleukin-1 in rheumatic diseases: its initial disappointments and recent successes in the treatment of autoinflammatory diseases. *Ann. N. Y. Acad. Sci.* **1182:** 111–123.

26. Stahl, N., A. Radin & S. Mellis. 2009. Rilonacept—CAPS and beyond: A scientific journey. *Ann. N. Y. Acad. Sci.* **1182:** 124–134.

27. Abuchowski, A., J.R. McCoy, N.C. Palczuk, T. van Es & F.F. Davis. 1977. Effect of covalent attachment of polyethylene glycol on immunogenicity and circulating life of bovine liver catalase. *J. Biol. Chem.* **252:** 3582–3586.

28. Steidler, L., P. Rottiers & B. Coulie. 2009. Actobiotics™ as a novel method for cytokine delivery: the interleukin-10 case. *Ann. N. Y. Acad. Sci.* **1182:** 135–145.

29. Steidler, L., W. Hans, L. Schotte, *et al*. 2000. Treatment of murine colitis by *Lactococcus lactis* secreting interleukin-10. *Science* **289:** 1352–1355.

30. Kozlowski, S., B. Cherney & R.P. Donnelly. 2009. Hurdles and leaps for protein therapeutics: cytokines and inflammation. *Ann. N. Y. Acad. Sci.* **1182:** 146–160.

# Pharmacologic Administration of Interleukin-2

## Inducing a Systemic Autophagic Syndrome?

**Antonio Romo de Vivar Chavez,**[a,b,f] **William Buchser,**[a]
**Per H. Basse,**[a,d] **Xiaoyan Liang,**[a,b,f] **Leonard J. Appleman,**[a,c]
**Jodi K. Maranchie,**[a,e] **Herbert Zeh,**[a,b] **Michael E. de Vera,**[a,b,f]
**and Michael T. Lotze**[a,b]

[a]*University of Pittsburgh Cancer Institute, Departments of* [b]*Surgery,* [c]*Medicine,*
[d]*Immunology, and* [e]*Urology,* [f]*Thomas E. Starzl Transplantation Institute, University of
Pittsburgh, Pittsburgh, Pennsylvania, USA*

The development of biologic therapies for patients with cancer has in part been impeded by the extraordinary complexity and intrinsic feedback mechanisms promoting homeostasis in tissue injury, repair, inflammation, and immunity. Recombinant interleukin 2 (IL-2) therapy was initiated in 1984 based on its role as the prototypic T-cell growth factor, with novel roles deduced late after its FDA approval in regulating not only effector T cells but also regulatory T cells. Complicating its application, even in the most sophisticated centers, has been the manageable but difficult toxicities attendant on its use in spite of clear evidence of complete responses in 5–10% of treated patients with melanoma and renal cell carcinoma with extraordinary durability lasting now for almost 25 years, thus tantamount to "cures." Although efforts have been made to diminish toxicity or enhance efficacy the only substantive advance in combination therapy has been the application of tumor-infiltrating lymphocytes and the antibody to CTLA4. A deeper understanding of the "limiting" toxicity associated with mild flu-like symptoms and more debilitating cytokine "storm" not forthcoming. Here we propose the notion that the systemic syndrome associated with IL-2 administration is due to global cytokine-induced autophagy and temporally limited tissue dysfunction. The possible role of autophagy inhibitors to enhance efficacy and limit toxicity as well as possible problems with this approach are considered.

*Key words*: interleuken-2; autophagy; HMGB1; chloroquine; biologic therapy; cancer

*For he (Chronos) learned from Earth and starry Heaven that he was destined to be overcome by his own son, strong though he was, through the contriving of great Zeus. Therefore he kept no blind outlook, but watched and swallowed down his children: and unceasing grief seized Rhea. Theogony, Hesiod*

## Introduction

Being overcome by cancer, one's own son as it were, is particularly concerning as the cells that give them origin are one's own. Autophagy, a critical "self-eating" cell process, serving as a vehicle to promote cell survival and clearance of effete organelles, denatured proteins, and pathogens, is initially a barrier to malignant transformation, but late in tumor growth promotes resistance to therapy. Cytokines, acting as paracrine and occasionally endocrine hormones, orchestrate both apoptosis and autophagy, and thereby the growth, function, and survival of cells of mesenchymal origin as well as the epithelia they support. Cytokines are best defined and are of particularly importance in the regulation of immunity. Clinically, cytokines have been the focus of substantial research in

Address for correspondence: Michael T. Lotze, lotzemt@upmc.edu

Cytokine Therapies: Ann. N.Y. Acad. Sci. 1182: 14–27 (2009).
doi: 10.1111/j.1749-6632.2009.05160.x © 2009 New York Academy of Sciences.

understanding the pathogenesis of several diseases and in their treatment.

During evolutionary history, signaling molecules first emerged in prokaryotes as a means to regulate cellular function. With the development of multicellular organisms, they have adopted alternative routes to reach their target cells. This is of particular importance when considering their use as treatment. For instance steroid and peptide hormones are produced by specialized endocrine organs. They have been used for decades as replacement strategies (adrenocorticoids, mineralocorticoids, insulin, thyroid hormone, etc.) or for the treatment of several pathologic conditions with a relatively low profile of side effects since their primary endocrine action is limited to the target tissue. This is not the case for cytokines, as these molecules are produced by individual mobile paracrine cells (e.g., macrophages or lymphocytes) or by tissue components (e.g., intestinal epithelium) and under normal conditions act locally (paracrine or autocrine) in the presence of infection or inflammation in individual tissues. Their concentration in the plasma is usually very low and its presence in the systemic circulation at higher than normal concentrations can cause a wide range of secondary desirable (pharmacologic) and undesirable (toxic) effects. In addition, most cytokines are pleiotropic, synergistic, and redundant, inducing many different biologic effects, depending on the target cells and the presence of other modulating factors. As redundant cytokines, they can exert very similar biologic effects.

This poses a challenge when using a cytokine as a systemic therapy because systemic effects are manifested on reaching the target organ or tissue with organ specific side effects. Treatment with interleukin-2 (IL-2) has an associated wide spectrum of side effects, ranging from a mild flu-like syndrome to a full-blown cytokine storm also associated with delivery of other immune active agents. This is also observed in the setting of multivisceral trauma and bacterial sepsis. IL-2 treatment is associated with objective tumor regression in about 20% of patients with

melanoma or renal cancer and half of these responses are complete. Furthermore 80% of these are ongoing beyond 10 years,[1] suggesting the ability of these therapies to truly evoke cures, albeit in a small percentage of patients. The wide range of side effects associated with IL-2 therapy keeps it from widespread used. Following a brief history of biologic therapy, we will focus on IL-2, its development as a therapeutic agent, and examine both its efficacy and toxicity through the lens of more modern biologic understanding.

## A Brief History of Biologic Therapy

Immunotherapy represents a relatively new form of cancer therapy. Biological therapy has however been used in other areas of medicine as early as the beginning of the 1800s. Edward Jenner utilized fluid from cattle infected with cowpox to inject humans and reported benefits in limiting smallpox contagion, leading him to develop the smallpox vaccine. The concept of immune memory was born. A century later, immune modifiers were used to treat cancer patients. William Coley, a New York surgeon, observed spontaneous tumor regression following bacterial, fungal, viral, and protozoal infections. This inspired Coley to work on the development of cancer immunotherapy, and he developed a killed bacterial vaccine for cancer in the late 1800s consisting of toxins derived from *Serratia marscescens* and *Streptococcus pyogenes*.

Since then several strategies to direct the immune response towards cancer cells and to understand the mechanisms of "immune escape" have been studied. Monoclonal antibodies are now the mainstay of cancer immunotherapy, and various forms of tumor vaccines have been extensively studied in the last decade as immune therapy, lately showing clear improvement in outcome when coupled with IL-2 therapy in patients with melanoma.[2] The use of cytokines has been also heavily studied in both preclinical and clinical studies. In the 1980s, the U.S. FDA approved the use of interferon for the treatment of patients with hairy-cell leukemia, chronic myelogenous leukemia, and Kaposi's sarcoma. The potent immunomodulatory effects and antitumor effects of IL-2 in the *in vitro* experiments and preclinical models prompted a rapid movement of IL-2 into the clinical setting.[3–5] In 1992 recombinant IL-2 received U.S. FDA approval for the treatment of patients with metastatic kidney cancer and several years later for patients with metastatic melanoma.

## Biology of IL-2

Over 30 years ago IL-2 was identified as a mitogenic molecule for T cells, produced by activated T cells, initially termed T cell growth factor.[6] It is now known to be one of several members of a family of cytokines interacting with and sharing the common IL-2R$\gamma_c$ chain, including IL-4, IL-7, IL-9, IL-15, and IL-21. IL-2 is produced mainly by activated T cells. In addition to promoting T-cell proliferation, IL-2 increases cytokine production and modifies the functional properties of B cells, NK cells, and possibly macrophages. Therefore IL-2 is critical in the activation of the adaptive immune response. Interestingly, IL-2 also plays an important role in limiting such responses and eliminating autoreactive T cells. IL-2 thus has a dual role that initiates immune responses but also limits their intensity and duration.

IL-2 is produced by activated T cells, although some reports of its production by murine but not human dendritic cells are quite convincing.[7] Resting T lymphocytes neither synthesize nor secrete IL-2 but can be induced to do so by antigen exposure along with costimulatory factors or by exposure to mitogenic stimuli. Although CD4$^+$ T-helper cells are the primary sources of IL-2, NK and CD8$^+$ T cells can also secrete it under certain conditions.[8,9]

IL-2 signals through a heterodimeric or trimeric high-affinity receptor complex consisting of subunits $\alpha\gamma$ or $\alpha\beta\gamma$, all of which have transmembrane domains. Only the $\beta$ and $\gamma$ chains bear intracytoplasmic domains that participate in signal transduction. The heterotrimer binds IL-2 with high affinity ($1.3 \times 10^{-11}$ $K_d$) and has a dissociation half-life of 50 minutes. The $\alpha$ chain alone (also called Tac or CD25) can bind IL-2 with low affinity but cannot signal. However, since the $\alpha$ chain is not expressed in resting T cells the $\beta/\gamma$ dimer binds IL-2 with intermediate affinity; activation of T cells by antigens or polyclonal mitogens leads to $\alpha$ subunit expression and formation of the high-affinity receptor trimer. The IL-2R $\gamma$ chain is also a component of the IL-4, IL-7, IL-9, and IL-15 receptor complexes, therefore all of these cytokines can act as T-cell growth factors.

The $\beta$ subunit of the IL-2 receptor has distinct cytoplasmic regions involved in signaling. A serine-rich region is required for induction of c-Myc protein; an acidic region mediates interaction with Lck and activation of the Ras pathway. Phosphorylation of this chain activates PI3 kinase. In addition, both the $\beta$ and $\gamma$ chains can interact with components of the Jak/Stat pathway. Mutations in either of these chains can lead to severe immune deficiencies in humans.

IL-2 has effects on non-T cells as well. Natural killer (NK) cells express IL-2R $\beta$-$\gamma$ dimers and thus they respond to IL-2 even in a resting state, being recruited into tissues through postcapillary venules where IL-2 is released. Relatively high concentrations of IL-2 are required for adequate signaling. However, following recruitment and stimulation with IL-2, NK cells, like T cells, begin to express the $\alpha$-chain, assembling the high-affinity trimers. IL-2-activated NK cells have enhanced cytolytic activity and secrete chemokines and cytokines, including IFN-$\gamma$, GM-CSF, and TNF-$\alpha$ that potently activate macrophages.

Activated or transformed B lymphocytes also express high-affinity IL-2R. IL-2 enhances proliferation and antibody secretion by normal B cells. The high-affinity trimer is found at about one third the density seen in activated T cells; the concentrations for activation are two- to threefold higher than are required to obtain T-cell responses. Human monocytes and macrophages constitutively express low levels of the IL-2Rβ chain, however on exposure to IL-2, IFN-γ, or other activating agents, they express high-affinity receptor trimers. Prolonged exposure to IL-2 improves the activated macrophage microbicidal and cytotoxic activities and promotes secretion of hydrogen peroxide, TNF-α, and IL-6. Neutrophils can be activated with high concentrations of IL-2; however in humans IL-2 treatment is associated with a decrease in neutrophil chemotaxis and FcR expression that confers susceptibility to infection.[10]

## Predictors of Prognosis and Response to Therapy

Since IL-2 was first introduced into the clinic for the treatment of patients with cancer, efforts have focused on trying to identify factors predictive of response, limiting the toxicity and expense for those who could not benefit. Previous studies suggest that there may be a subgroup of patients with advanced cancer who will benefit from this form of immunotherapy.

To date, several factors have been identified as predictors of response to therapy including performance status, histological subtype of renal cell carcinoma (RCC) or metastatic site (lung, cutaneous, and nodal being better sites). Table 1 summarizes both prognostic factors and predictors of response to IL-2 based therapy gleaned from several clinical trials and retrospective studies. We have listed both clinical parameters and molecular markers. In general good prognostic factors correlate with a better response to IL-2 therapy. Whether these factors directly influence IL-2 responses per se or are just indicators of overall survival is not clear. It remains largely unknown whether adding the assessment of molecular markers to the conventional clinical parameters can contribute to enhanced accuracy in predicting responses to high-dose IL-2.

Carbonic anhydrase IX (CAIX) is the most significant molecular marker described in kidney cancer to date. Decreased CAIX levels are independently associated with poor survival in advanced RCC. This marker could be used to predict clinical outcome and identify high-risk patients. CAIX has been identified as independent factor that predicts response of RCC to high-dose IL-2.[11]

Identifying patients likely to respond and avoiding treatment for those who whom are least likely to respond could limit the number of patients subjected to IL-2 related toxicity. Sadly, it doesn't broaden the number of patients that could actually benefit from this therapy. Reducing toxicity without significantly compromising or even improving efficacy would have the greatest impact in the use of high-dose IL-2 for the treatment of patients with cancer.

## Understanding IL-2 Toxicity

The mechanisms responsible for high-dose IL-2 responses in patients with RCC and melanoma are not clear, and more than one mechanism may be involved[12] and include the relative roles of IL-2 activated NK cells and T cells. The mechanism for IL-2 toxicity is also poorly understood. Cytokine storm has been proposed as responsible for the side effects associated with IL-2 treatment. Identification of serum secondary cytokines such as TNF-α, IL-1 and IL-6[13] and many others has been found. Attempts to neutralize the effect of such cytokines by co-administration of soluble receptors for TNF, soluble IL-1R antagonists or inhibitors have yielded only modest reduction of the serious side effects associated with high dose IL-2 treatment.[14–16] It is crucial to understand the fundamental mechanism of effective IL-2 therapy and dissociate these mechanisms from the mediators of toxicity.

Attempts to alter the side effect profile of IL-2 have resulted in diminished toxicity but failed to enhance and often limited the efficacy of high dose IL-2 treatment. A continuous infusion of IL-2 over a 5-day period for the treatment of patients with advanced cancer, delivered only

**TABLE 1.** Predictors of IL-2 Clinical Response

| | Number of patients | Survival | | IL-2 response | | Ref. |
|---|---|---|---|---|---|---|
| | | Melanoma | RCC | Melanoma | RCC | |
| **Histology** | | | | | | |
| *Carbonic anhydrase IX (CAIX)* | 122 | NR | + | | + | 11,42,43 |
| *Ki-67 expresion in nephrectomy specimen* | 40 | | − | | − | 44 |
| *Bcl-2 and Fas in primary lession* | 40 | | − | | − | 45 |
| *Clear cell subtype* | 231 | NR | + | NR | + | 46 |
| *Uveal melanoma* | $7^{47},6^{48}$ | − | | − | | 47,48 |
| **Clinical Predictors** | | | | | | |
| *Metastasis at diagnosis* | $40^{44}374$ mm$^{49}$ | − | − | − | − | 44,49 |
| *Good performance status* | 110 | + | + | | | 50 |
| *Prior nephrectomy* | | | + | | | |
| *DFI longer than 12 months* | | + | + | | | |
| *Bone disease* | | − | − | | | |
| *Low number of metastatic sites* | | + | + | | | |
| *Immunologic side effects (e.g. vitiligo)* | 374 | | | + | | 49 |
| *Previously failed other immunotherapies* | 509 | | − | | | 51 |
| **Blood Biomarkers** | | | | | | |
| *VEGF* | 49 | − | − | − | − | 52 |
| *Fibronectin* | | − | − | − | − | |
| *Low/normal C-reactive protein* | $110^{50}, 81^{53}$ | + | + | + | | 50,53 |
| *Normal albumin* | 110 | | + | | | 50 |
| *Low/normal fibrinogen* | | | + | | | |
| *Low/normal lactate dehydrogenase* | | | + | | | |
| Circulating mature (CD11c)$^+$ DCs | 25 | | + | | + | 54 |
| IL-12 | | | + | | + | |
| Circulating Tregs (CD4$^+$CD25hi) | 45MM/12RCC | − | − | | | 55 |
| Decrease in Circulating Tregs (CD4$^+$CD25hi) | 45MM/12RCC | + | + | + | + | |
| Lymphocytosis after therapy | 374 | + | | + | | 49 |

(+) and (−) indicate a positive or negative correlation; NR, not relevant; DFI, disease-free interval; VEGF, J39 endothelial growth factor; IL, interleukin; RCC, renal cell carcinoma; M, melanoma.

Variables in *italics* indicate pretreatment parameters.

a fifth of the IL-2 used in the high-dose intravenous bolus regimen. Although this regimen produced response rates similar to those of high-dose IL-2 administration, the toxicity was also comparable.[17]

In the case of patients with metastatic melanoma, the response rates for lower dose regimens have been consistently inferior to high-dose IL-2.[18] For renal cancer, a phase 3 randomized clinical trial compared high-dose IL-2 with either a subcutaneous or IV low-dose IL-2 revealed response rates of 10% and 13%, respectively, compared to 21% for the high-dose IL-2 limb, consistent with the findings in patients with melanoma.

# IL-2 Toxicity Compared with Flu-like Syndrome and Septic Shock

Acute systemic inflammatory conditions are characterized by detectable elevation of immune mediators. These mediators of inflammatory processes are often produced in response to organisms at the site of infection and have potent paracrine and endocrine effects. These same mediators can, at low levels, induce a flu-like syndrome and at higher levels damage distant organs with an exaggerated systemic inflammatory response syndrome (SIRS). Bone et al. defined four sepsis-related clinical syndromes that caused organ damage with activation of endogenous inflammatory responses[19] Such syndromes are caused in large part by mediators released from lymphocytes, macrophages, granulocytes, and endothelial cells including cytokines, enzymes, and reactive oxygen species (ROS). These molecules have a beneficial role in limiting invading microorganisms, but as immune mediators they also initiate coagulation cascades, potentiate the secretion of additional cytokines and vasoactive agents, and increase capillary membrane permeability. This can result in organ dysfunction manifested by lactic acidosis, oliguria, and a depressed level of consciousness. Resolving hypotension in such patients is usually associated with a corresponding improvement in organ function. The most severe category of SIRS is septic shock in which hypotension is often refractory to fluid resuscitation and application of vasoactive agents. Despite deeper understanding of the clinical manifestations of sepsis, trials of agents directed at altering the host's response have had limited results, and it appears that several additional factors may alter the efficacy of these agents. To date, no single mediator-specific anti-inflammatory agent has been shown to significantly improve survival.[20]

The effects of IL-2 toxicity have been compared to that of septic shock in both clinical manifestations and the profile of elevated inflammatory mediators such as IFN-γ, TNF, and IL-1. Although several studies suggest that a direct leukocyte-mediated endothelial injury plays a dominant role in the vascular changes and organ dysfunction during IL-2 toxicity. The rapid reversal of IL-2 mediated toxicity with few if any observable long-term sequelae limit enthusiasm for this interpretation. In fact, even with similar cytokine profiles, Il-2 toxicity seems to be independent of such elevations in IFN-γ and TNF in a murine model of IL-2/IL-12 toxicity. NK cells alone were able to cause comparable toxicity in SCID mice that lack both T and B lymphocytes.[21] Furthermore, inhibition of NF-κB signaling *in vivo* is ineffective in preventing mortality in this model in contrast

with its protective effect in mice administered lethal amounts of endotoxin.[22] The similarities between Gram-negative sepsis and the hemodynamic changes induced by high-dose IL-2 are prominent. However, the clinical course of these two entities is also quite different.

The overall risk for death in patients with sepsis is 16%, while risk for death in patients with severe sepsis or septic shock are 20% and greater than 46%, respectively.[23] In contrast, no toxic-related deaths secondary to high-dose IL-2 were reported in six recent randomized clinical trials of IL-2 treatment for RCC.[24] Sepsis-related organ dysfunction often progresses to organ failure. However IL-2-related organ dysfunction rarely progresses to irreversible organ damage. In a postmortem study that we (MTL) conducted, the pathologic findings of 19 cancer patients who died during or soon after receiving IL-2 therapy where analyzed. Most of the patients had widespread metastatic disease at the time of death. However analysis of metastasis free organs provided interesting information. The cardiac pathologic findings were heterogeneous. Only a small number of samples showed lymphoctytic myocarditis, myocytolysis, fibrosis, or band necrosis. No statistically significant correlation between postmortem cardiac findings and clinical cardiac events in these patients was found. Six out of 19 renal pathology specimens showed lymphocytic interstitial infiltrates; but no correlation was detected with clinical nephrotoxicity. Analyses of lung and liver were unfortunately confounded by the high incidence of metastatic disease to these organs (e.g., 18 out of 19 in lung specimens) and the lymphocytic infiltrates within and adjacent to the tumor.[25]

Despite the similar manifestations of IL-2-related toxicity and severe forms of sepsis, the progression of these two different entities seems to be different: sepsis-related manifestations are often refractory to treatment and frequently lead to permanent organ damage whereas IL-2 related toxicity is highly predictable, reversible, and only rarely causes permanent organ failure.

## Systemic Autophagic Syndrome

We believe that the difference in outcomes following sepsis and systemic IL-2 toxicity relies at the cellular level with the cellular pathway of cell survival called autophagy, specifically playing the dominant role.

Autophagy is a highly regulated catabolic process involving the degradation of the cell's own components. The formation of a double or triple membrane vesicle, the autophagosome, around organelles and/or proteins initiates the process. This vesicle travels along micrutubules or actin cytoskeletal elements to fuse with lysosomes in a perinuclear location, forming a phagolysosome. The contents are then degraded to be utilized as energy substrates for the cell. It is a process that plays a part in normal cell growth, development, and homeostasis, helping to maintain a balance between the synthesis, degradation, and subsequent recycling of cellular products. It is also a major mechanism by which a starving cell reallocates nutrients from nonessential processes. More importantly, it represents a path of cell survival that is potentially reversible. Cells with enhanced stress may augment autophagy and survive at the expense of compromising the normal function of the cell.

Cells have detectable baseline autophagy under normal conditions. Additional autophagy is induced by autophagic stimuli such as proteosome inhibition, genomic or ER stress, or oxidant stress. The movement of autophagosomes from the periphery towards the nucleus to fuse with the lysosome is known as autophagic flux and varies in magnitude in different cell types and under various physiologic and pathologic conditions. Autophagic flux in a stressed cell decreases when the cell has returned to a non-stressed condition[26] (e.g., from a glucose-deprived to a glucose-rich environment). Thus, if specific manifestations of IL-2 toxicity may be related to increased autophagic flux in individual tissues, as we hypothesize, the reversible nature of enhanced autophagic flux in a cell may explain the rather rapid decrease in high-dose IL-2 toxicity when treatment is withdrawn. Moreover, we hypothesize that some of the specific manifestations of IL-2 toxicity may be related to increased autophagic flux in individual tissues.

Increased autophagic flux can be detrimental to cardiac myocytes under certain conditions both *in vivo* and *in vitro*.[27] Inhibition of autophagy may be mediated through downregulation of an autophagy-related gene (ATG-6), Beclin 1, required for the formation of the autophagosome. This change is protective during ischemia-reperfusion cardiac injury. Inhibition of autophagy by Beclin 1 knockdown also increases cell viability in response to oxidant injury mediated by $H_2O_2$ in cardiac myocytes *in vitro*.

During high-dose IL-2 treatment there is a dramatic fall in serum albumin that is generally attributed to the egress of albumin containing fluid from the circulation to the extravascular space. However hepatic protein synthesis may also be suppressed during IL-2 treatment as part of the "acute phase" response. Supernatants of IL-2 primed peripheral blood mononuclear cells increase the production of acute phase reactants and almost totally suppress that of albumin in an hepatocellular carcinoma cell line.[28] This dramatic decrease in albumin may be explained by an increased autophagic flux and excessive protein turnover in hepatocytes. Many patients have transient prolongation of prothrombin time without evidence of consumptive coagulopathy, which also suggests suppressed protein output by the liver. Autophagy does not stop protein synthesis in the cell. Indeed, one of the primary functions of autophagy is to sustain synthesis of essential proteins including autophagy-execution proteins. Nonetheless certain stress stimuli that induce autophagy, such as starvation, markedly attenuate general protein synthesis. This process ensures that the cell has sufficient amino acids to synthesize the proteins that are essential for its survival.[29]

For many years, it was thought that autophagy-inducing therapies would kill cells through autophagy (i.e., induced "autophagic cell death"). However, recent evidence suggests that specific inhibition of autophagy with siR-NAs targeted against ATG genes actually increases cell death,[30] indicating that autophagy is, as we like to call it, a "programmed cell survival" pathway which represents an attempt to cope with cellular stress. In the setting of cancer treatment, stress induced by cytotoxic agents, is associated with marked autophagy in the surviving cell fraction. As most cytotoxic agents enhance autophagy, this suggests that simultaneous inhibition of autophagy might enhance the efficacy of existing cancer treatments. Indeed, in mice harboring c-Myc-induced lymphomas the drug chloroquine, an antimalarial drug with alkalinizing lysosomotropic properties that impairs autophagic degradation, enhances the ability of either p53 activation or a DNA alkylating agent to induce tumor cell death and tumor regression. Application of ATG5 siRNA showed similar results.[31] Several studies have suggested that autophagy may act as a protective mechanism in tumor cells in which cell death is induced by drugs, and that inhibition of autophagy provides antitumor effects alone[32–34] or synergistic effect with such drugs.[35]

Furthermore, a recent study suggests that the ability to initiate autophagy in response to stress is enhanced in oncogenic transformed cells, reflecting a stronger dependence of transformed cells on autophagy for survival. A combined use of chloroquine and a proteasome inhibitor (bortezomib) suppressed tumor growth more significantly than either agent alone. This also suggests that autophagy inhibition may target transformed cells while relatively sparing non-transformed cells.[36] This and other evidence suggesting a potential prosurvival and pro-tumorigenic role for autophagy, supports our interest in combining antiautophagic strategies with high-dose IL-2. Chloroquine however has effects on both tumor cells and immune cells.[37] Whether the antitumor effects

of chloroquine may be explained by its anti-autophagic effects alone remains to be seen. We are currently exploring the effects of combined chloroquine and IL-2 treatment in murine experiments.

## After Death, What? The Role of DAMPs

Autophagy occurs when cells degrade their constituents. A baseline level of autophagy occurs probably in most normal cells to prevent the accumulation of protein aggregates and defective cellular structures. Additionally under certain stressful stimuli such as glucose starvation, high temperature, low oxygen, or intracellular stress including accumulation of damaged organelles or denatured proteins, microbial invasion, etc., the inducible autophagy signaling pathways are activated. Therefore it is not surprising that observable autophagy is a frequent occurrence and concomitant in dying cells. However we now know that the death of a cell is not as simple as we once thought it was. How the cell dies and what happens immediately thereafter is of great importance in modern biologic therapy.

Cell death can be classified according to its morphological criteria as apoptotic, or necrotic. However these types of cell death have biological differences in terms of signaling pathway activation, functional aspects (programmed versus accidental) and more importantly immunogenic or nonimmunogenic cell death.

Unlike necrosis and apoptosis that are forms of cell death, autophagy commonly contributes to cell survival. However autophagic cell death can take place when cellular apoptotic machinery is crippled or when the autophagic flux is induced to such an extent that the cells literally eat themselves to death. Both circumstances are present in the setting of cancer wherein the conditions are such that autophagy is always induced (low oxygen and nutrients due to poor blood supply, mutant proteins, chemotherapeutic agents, etc.) and in cells populations in which mutations and Darwinian selection confers resistance to apoptotic cell death.[29]

The relationship between autophagy and apoptosis is complex and the specific molecular pathways are beyond the scope of this review. Although not mutually exclusive, under many circumstances there seems to be a polarization between these two catabolic pathways. Autophagy and apoptosis share many common inducers, and in many cases the process that dominates has a cross-inhibitory interaction with the other. In other words, situations that stimulate apoptosis inhibit autophagy and increased apoptosis inhibits autophagy. For example, autophagy can inactivate caspases that are involved in many steps of the apoptotic pathway. In the case of nutrient deprivation, autophagic flux is increased, creating a metabolic state with increased adenosine triphosphate (ATP) production which itself is antiapoptotic. On the other hand, highly apoptotic stimuli such as DNA damage and activation of p53 initiate the apoptotic pathway and caspase cascades which is a rapid self-amplifying process that precludes full manifestation of a simultaneous autophagic response.[30] Consistent with this, autophagy inhibitors such as chloroquine inhibit colon cancer cell growth by induction of apoptosis *in vitro* and *in vivo*, prolonging survival in murine models.[32] We have confirmed these findings in murine pancreatic tumor models.

Another major difference in autophagic and apoptotic cell death is their immunogenicity, mainly dictated by molecules released by the cell after death. These molecules are termed damage associated molecular pattern (DAMP) molecules. The best characterized DAMP, high mobility group box protein-1 (HMGB1) is a highly evolutionary conserved molecule abundant in the nucleus of normal cells where it binds and bends DNA and serves as a regulator of transcription. When released outside the cell it acts as a cytokine/inflammatory mediator, which we have defined as an "epicrine" medi-

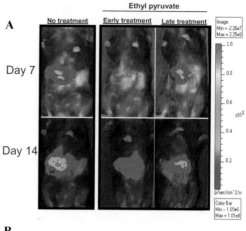

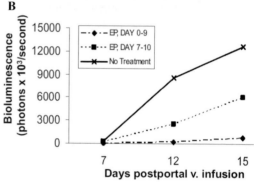

**Figure 1.** EP inhibits hepatic tumor growth. Portal vein injection of MC38 tumor cells in C57BL/6 mice, which received no treatment or EP (80 mg/kg) administered i.p. from Days 0 to 9 (early treatment) or from Days 7 to 10 (late treatment). Luciferase-labeled MC38 tumor cells infused intraportally allowed *in vivo* tracking of tumor growth with bioluminescence intensity (**A**). Data are representative of three individual experiments. (**B**) Graphic representation of bioluminescence readings for the three different groups. (From Ref. 39, with permission)

ator to distinguish it from the primary process acting on the genes, epigenetic, and the purposeful release of paracrine or autocrine agents. HMGB1 can be released into the extracellular space under two conditions: (1) it is acetylated and ADP ribosylated and actively secreted by activated macrophages or other immune cells and (2) it is released by cells undergoing nonapoptotic cell death or in response to stress and autophagy.[38] The function of HMGB1 has been extensively studied in both its intranuclear and extracellular forms, but increasing

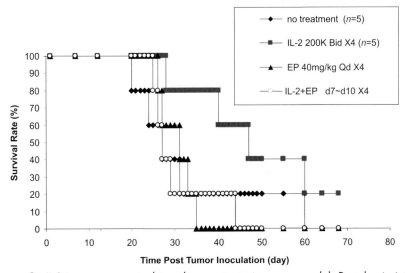

**Figure 2.** IL-2 increases survival in a liver metastasis mouse model. Portal vein injection of MC38 tumor cells in C57BL/6 mice, which received either no treatment, IL-2 200 K IU/kg Bid (i.p.), ethyl pyruvate (EP) (40 mg/kg) administered i.p. from Days 7 to 9 or IL-2 plus EP as described. Survival curves are shown.

evidence suggests that HMGB1 has important biological functions when released from the nucleus to the cytoplasm.

Measures of autophagy have been classically performed by demonstration of autophagic vesicles. Phosphatidylethanolamine (PE) is a lipid component of cellular membranes and autophagic vesicles. PE-conjugation of LC3-I protein (the soluble form) during autophagy results in a nonsoluble form of the complex or LC3-II (lipidated form) that stably associates with the autophagosomal membrane. Thus, autophagy can be detected microscopically by observing the localization pattern (perinuclear LC3 spots) of fluorescently tagged LC3, best visualized by fluorescence microscopy or, as we have shown, by imaging cytometry.

Recently, we demonstrated that cytosolic expression of HMGB1 is a critical regulator of sustained autophagy. One of the HMGB1 receptors, the receptor for Advanced Glycation End products (RAGE) appears to be similarly critical for sustained autophagy. Stimuli that enhance reactive oxygen species promote translocation of HMGB1 to the cytosol where it interacts with Beclin 1 and thereby enhances autophagic flux. Furthermore a mutation in cysteine 106 promotes cytosolic localization of HMGB1 and promotes sustained autophagy.[39]

Drugs that inhibit HMGB1 cytoplasmic translocation such as ethyl pyruvate (EP) can limit starvation-induced autophagy. EP can suppress liver tumor growth significantly in a dose-dependent manner (Fig. 1). Serum HMGB1 decreased significantly after EP administration. *In vitro* treatment with EP showed increases in LC3-II and cleaved PARP, markers of autophagy and apoptosis respectively.[40] This increase in autophagy and apoptosis suggest increased cellular stress. Other mechanisms by which EP exerts its antitumor effects are likely taking place and are yet to be investigated.

The possible synergistic effects of EP and IL-2 were also investigated by our group showing no added benefit from EP compared with IL-2 treatment alone (Fig. 2) High dose IL-2 increased the survival time significantly. However late treatment with EP (as opposed to early treatment in Fig. 1) alone or in combination with IL-2 did not show significant differences when compared with the untreated group.

Dying cells in which autophagy is elevated release HMGB1 into the extracellular space.[41] Since the presence of HMGB1 and its oxidation state regulate immunity or tolerance, regulation of autophagy could also control the immunogenicity of a dying cell. We believe that HMGB1 is a critical proautophagic protein that enhances cell survival and limits programmed apoptotic death. We have developed novel imaging cytometric measures of autophagy which can be coupled with measures of p62 and LC3 punctae in cell culture.

Increased autophagic flux was observed in renal carcinoma cells co-cultured with peripheral blood lymphocytes (PBL) compared to cancer cells alone. Adding IL-2 to the cultures resulted in increased effector-mediated killing of tumor cells. Furthermore, when analyzing the remaining cells, the amount of LC3 spots is higher on a per cell basis indicating a possible contribution of IL-2 to the capacity of effector cells to also induce autophagy (Fig. 3). The role of effector cells as inducers of cell survival is intriguing and is the focus of current investigations ongoing by our group.

In an *in vivo* experiment we treated mice with high-dose IL-2 and found evidence of HMGB1 translocation from the nucleolus to the cytosol. This correlates with the finding of increased autophagic flux as measured by LC3 punctae (Fig. 4). Our preliminary data support the notion that IL-2 administration increases autophagic flux in parenchymal cells. Whether increased autophagy impairs liver function and other similarly reversible clinical manifestations remains to be seen.

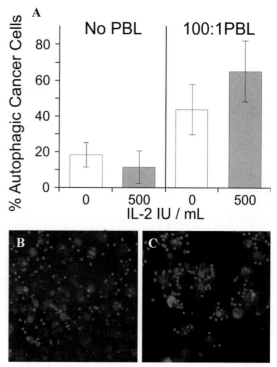

**Figure 3.** Increased autophagy in renal cancer cell line following exposure to NK cells and IL-2. Cells from RCC4 renal carcinoma cell line were plated at 12,000 cells/well in a 96-well plate and incubated with peripheral blood lymphocytes (PBL) 16h. (**A**) Average percent of autophagic cells (distinguished as having 8 or more LC3 punctae ± SD). When no effectors were present, IL-2 was unable to induce autophagy. With 100:1 effector to target ratio of PBLs, 43.4% of the remaining cancer cells had autophagy markers, while 64.7% were autophagic after addition of IL-2. Averages were based on fields per condition. (**B,C**) Fluorescent images of co-cultures stained for Hoechst (blue) and LC3 (red). RCC4 cells (large diffuse Hoechst signal) have minimal basal autophagy (**B**), which is increased in the presence of high IL-2 (**C**).

## Conclusions

Although high-dose IL-2 has demonstrated promising results and durable complete responses in some patients with melanoma and renal cell carcinoma, the spectrum of potentially life-threatening side effects has limited its use to a highly selected small number of patients. The efficacy of the high-dose treatment, unmatched by low-dose regimens, brings a necessity to dissect the therapeutic components of IL-2 treatment from those of toxicity. We need to move beyond the current clinical and cellular markers onto more specific indicators of the immune response. A recent publication showed that the effects of IL-2 in combination with antiretroviral (ARV) therapy in patients with human immunodeficiency virus (HIV) greatly

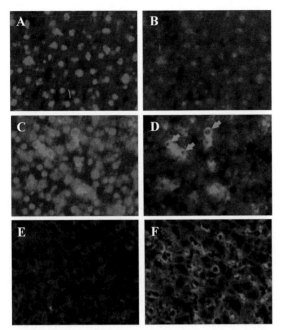

**Figure 4.** IL-2 treatment induces HMGB1 translocation to the cytosol in hepatic cells. C57B16 female mouse were treated with 200 K IU i.p. injections of IL-2 or PBS twice a day during 5 days. The livers were harvested and stained with Hoechst, to visualize by fluorescence microscopy the nuclei (blue) and with HMGB1 AB (red). (**A**) and (**B**) are specimens from the non-treatment group; (**C**) and (**D**) specimens from the IL-2 treatment group. Hoechst has been edited out to visualize the HMGB1 localization (**B**) or its absence (**D**; blue arrows) in the nucleus. Translocation of HMGB1 to the cytosol can be seen in the IL-2 treatment group and almost complete absence in the nuclear area. (**E**) and (**F**) are sections of the liver stained for LC3 for the nontreatment group and IL-2 treatment respectively. Markedly increased LC3 staining can be seen in the IL-2 treatment group consistent with increased autophagic flux.

improves their CD4$^+$ cell count compared with ARV alone. The CD4$^+$ cell count remains the best single indicator of immunodeficiency related to HIV infection. However this combined therapy yielded no clinical benefit. Furthermore, patients with higher baseline CD4$^+$ cell counts who showed the greatest expansions of CD4$^+$ T cells also had higher risk of deleterious effects of IL-2 therapy and a higher risk of death.[56] The exact mechanism of action of IL-2 treatment is not well understood, and the current understanding of its toxicity has yielded little progress in its reduction. We propose here a substantially different notion, based in our observations both in the clinical setting, and preliminary experimental data. We suggest that a "systemic autophagic syndrome" is the biologic basis for the reversible toxic manifestations of this therapy. Ongoing studies from our group will focus on finding ways to counteract these symptoms and to determine whether inhibiting autophagy may enhance high-dose IL-2 efficacy and limit its toxicity.

## Acknowledgments

We appreciate discussions with Drs. Steven Rosenberg, Michael Atkins, Patrick Hwu, Craig Thompson, and Ravi Amaravadi from Penn and Dr. Eileen White from UMDNJ over the last year in consideration of the critical role of autophagy and available inhibitors for cancer therapy. This project was funded by a grant from the NIH 1 P01 CA 101944-04 (Michael T. Lotze) Integrating NK and DC into Cancer Therapy National Cancer Institute; and a grant with the Pennsylvania Department of Health. The Department specifically disclaims responsibility for any analyses, interpretations or conclusions derived from this work.

## Conflicts of interest

The authors declare no conflicts of interest.

## References

1. Rosenberg, S.A., J.C. Yang, S.L. Topalian, *et al.* 1994. Treatment of 283 consecutive patients with metastatic melanoma or renal cell cancer using high-dose bolus interleukin 2. *JAMA* **271:** 907–913.
2. Schwartzentruber, D.J. *et al.* 2009. A phase III multi-institutional randomized study of immunization with the gp100: 209-217(210M) peptide followed by high-dose IL-2 compared with high-dose IL-2 alone in patients with metastatic melanoma. *J. Clin. Onc.* **27:**185, part 2/2, 807S.
3. Lotze, M.T., L.W. Frana, S.O. Sharrow, *et al.* 1985. In vivo administration of purified human interleukin 2.

I. Half-life and immunologic effects of the Jurkat cell line-derived interleukin 2. *J. Immunol.* **134:** 157–166.

4. Lotze, M.T., Y.L. Matory, S.E. Ettinghausen, *et al.* 1985. In vivo administration of purified human interleukin 2. II. Half life, immunologic effects, and expansion of peripheral lymphoid cells in vivo with recombinant IL 2. *J. Immunol.* **135:** 2865–2875.

5. Lotze, M.T., R.J. Robb, S.O. Sharrow, *et al.* 1984. Systemic administration of interleukin-2 in humans. *J. Biol. Response Mod.* **3:** 475–482.

6. Gordon, J. & L.D. MacLean. 1965. A lymphocyte-stimulating factor produced in vitro. *Nature* **208:** 795–796.

7. Granucci, F., I. Zanoni, N. Pavelka, *et al.* 2004. A contribution of mouse dendritic cell-derived IL-2 for NK cell activation. *J. Exp. Med.* **200:** 287–295.

8. Cousens, L.P., J.S. Orange & C.A. Biron. 1995. Endogenous IL-2 contributes to T cell expansion and IFN-gamma production during lymphocytic choriomeningitis virus infection. *J. Immunol.* **155:** 5690–5699.

9. Schmidt, R.E., T. Hercend, D.A. Fox, *et al.* 1985. The role of interleukin 2 and T11 E rosette antigen in activation and proliferation of human NK clones. *J. Immunol.* **135:** 672–678.

10. Klempner, M.S., R. Noring, J.W. Mier & M.B. Atkins. 1990. An acquired chemotactic defect in neutrophils from patients receiving interleukin-2 immunotherapy. *N. Engl. J. Med.* **322:** 959–965.

11. Phuoc, N.B., H. Ehara, T. Gotoh, *et al.* 2008. Prognostic value of the co-expression of carbonic anhydrase IX and vascular endothelial growth factor in patients with clear cell renal cell carcinoma. *Oncol. Rep.* **20:** 525–530.

12. Romo de Vivar Chavez, A., M.E. de Vera, X. Liang & M.T. Lotze. 2009. The biology of interleukin-2 efficacy in the treatment of patients with renal cell carcinoma. *Med. Oncol.* **26**(Suppl 1): 3–12.

13. Gemlo, B.T., M.A. Palladino, Jr., H.S. Jaffe, *et al.* 1988. Circulating cytokines in patients with metastatic cancer treated with recombinant interleukin 2 and lymphokine-activated killer cells. *Cancer Res.* **48:** 5864–5867.

14. Atkins, M.B., B. Redman, J. Mier, *et al.* 2001. A phase I study of CNI-1493, an inhibitor of cytokine release, in combination with high-dose interleukin-2 in patients with renal cancer and melanoma. *Clin. Cancer Res.* **7:** 486–492.

15. Du Bois, J.S., E.G. Trehu, J.W. Mier, *et al.* 1997. Randomized placebo-controlled clinical trial of high-dose interleukin-2 in combination with a soluble p75 tumor necrosis factor receptor immunoglobulin G chimera in patients with advanced melanoma and renal cell carcinoma. *J. Clin. Oncol.* **15:** 1052–1062.

16. McDermott, D.F., E.G. Trehu, J.W. Mier, *et al.* 1998. A two-part phase I trial of high-dose interleukin 2 in combination with soluble (Chinese hamster ovary) interleukin 1 receptor. *Clin. Cancer Res.* **4:** 1203–1213.

17. West, W.H., K.W. Tauer, J.R. Yannelli, *et al.* 1987. Constant-infusion recombinant interleukin-2 in adoptive immunotherapy of advanced cancer. *N. Engl. J. Med.* **316:** 898–905.

18. DeVita, V.T., S. Hellman & S.A. Rosenberg. 1995. *Biologic Therapy of Cancer*, 2nd edn. Lippincott. Philadelphia.

19. Bone, R.C., R.A. Balk, F.B. Cerra, *et al.* 1992. Definitions for sepsis and organ failure and guidelines for the use of innovative therapies in sepsis. The ACCP/SCCM Consensus Conference Committee. American College of Chest Physicians/Society of Critical Care Medicine. *Chest* **101:** 1644–1655.

20. Deans, K.J., M. Haley, C. Natanson, *et al.* 2005. Novel therapies for sepsis: a review. *J. Trauma.* **58:** 867–874.

21. Carson, W.E., H. Yu, J. Dierksheide, *et al.* 1999. A fatal cytokine-induced systemic inflammatory response reveals a critical role for NK cells. *J. Immunol.* **162:** 4943–4951.

22. Bohrer, H., F. Qiu, T. Zimmermann, *et al.* 1997. Role of NFkappaB in the mortality of sepsis. *J. Clin. Invest.* **100:** 972–985.

23. Rangel-Frausto, M.S., D. Pittet, M. Costigan, *et al.* 1995. The natural history of the systemic inflammatory response syndrome (SIRS). A prospective study. *JAMA* **273:** 117–123.

24. Hotte, S., T. Waldron, C. Canil & E. Winquist. 2007. Interleukin-2 in the treatment of unresectable or metastatic renal cell cancer: a systematic review and practice guideline. *Can. Urol. Assoc. J.* **1:** 27–38.

25. Kragel, A.H., W.D. Travis, L. Feinberg, *et al.* 1990. Pathologic findings associated with interleukin-2-based immunotherapy for cancer: a postmortem study of 19 patients. *Hum. Pathol.* **21:** 493–502.

26. Boya, P., R.A. Gonzalez-Polo, N. Casares, *et al.* 2005. Inhibition of macroautophagy triggers apoptosis. *Mol. Cell Biol.* **25:** 1025–1040.

27. Matsui, Y., H. Takagi, X. Qu, *et al.* 2007. Distinct roles of autophagy in the heart during ischemia and reperfusion: roles of AMP-activated protein kinase and Beclin 1 in mediating autophagy. *Circ. Res.* **100:** 914–922.

28. Mier, J.W., C.A. Dinarello, M.B. Atkins, *et al.* 1987. Regulation of hepatic acute phase protein synthesis by products of interleukin 2 (IL 2)-stimulated human peripheral blood mononuclear cells. *J. Immunol.* **139:** 1268–1272.

29. Levine, B. 2007. Cell biology: autophagy and cancer. *Nature* **446:** 745–747.

30. Maiuri, M.C., E. Zalckvar, A. Kimchi & G. Kroemer. 2007. Self-eating and self-killing: crosstalk between autophagy and apoptosis. *Nat. Rev. Mol. Cell Biol.* **8:** 741–752.

31. Amaravadi, R.K., D. Yu, J.J. Lum, *et al.* 2007. Autophagy inhibition enhances therapy-induced apoptosis in a Myc-induced model of lymphoma. *J. Clin. Invest.* **117:** 326–336.

32. Zheng, Y., Y.L. Zhao, X. Deng, *et al.* 2009. Chloroquine inhibits colon cancer cell growth in vitro and tumor growth in vivo via induction of apoptosis. *Cancer Invest.* **27:** 286–292.

33. Maclean, K.H., F.C. Dorsey, J.L. Cleveland & M.B. Kastan. 2008. Targeting lysosomal degradation induces p53-dependent cell death and prevents cancer in mouse models of lymphomagenesis. *J. Clin. Invest.* **118:** 79–88.

34. Dang, C.V. 2008. Antimalarial therapy prevents Myc-induced lymphoma. *J. Clin. Invest.* **118:** 15–17.

35. Bellodi, C., M.R. Lidonnici, A. Hamilton, *et al.* 2009. Targeting autophagy potentiates tyrosine kinase inhibitor-induced cell death in Philadelphia chromosome-positive cells, including primary CML stem cells. *J. Clin. Invest.* **119:** 1109–1123.

36. Ding, W.X., H.M. Ni, W. Gao, *et al.* 2009. Oncogenic transformation confers a selective susceptibility to the combined suppression of the proteasome and autophagy. *Mol. Cancer Ther.* **8:** 2036–2045.

37. Apetoh, L., F. Ghiringhelli, A. Tesniere, *et al.* 2007. Toll-like receptor 4-dependent contribution of the immune system to anticancer chemotherapy and radiotherapy. *Nat. Med.* **13:** 1050–1059.

38. Lotze, M.T., H.J. Zeh, A. Rubartelli, *et al.* 2007. The grateful dead: damage-associated molecular pattern molecules and reduction/oxidation regulate immunity. *Immunol. Rev.* **220:** 60–81.

39. Daolin Tang, R.K., P. Loughran, A.M. Farkas, *et al.* 2008. The damage associated molecular pattern molecule (DAMP) high mobility group box protein-1 (HMGB1) is an activator of autophagy. *J. Leuk. Biol.* **84:** A19–20. (Abstract).

40. Liang, X., A. Romo-Vivar, N.E. Schapiro, *et al.* 2009. Ethyl pyruvate administration inhibits hepatic tumor growth. *J. Leukoc. Biol.* 2009 Jul 7.

41. Thorburn, J., H. Horita, J. Redzic, *et al.* 2009. Autophagy regulates selective HMGB1 release in tumor cells that are destined to die. *Cell Death Differ.* **16:** 175–183.

42. Atkins, M., M. Regan, D. McDermott, *et al.* 2005. Carbonic anhydrase IX expression predicts outcome of interleukin 2 therapy for renal cancer. *Clin. Cancer Res.* **11:** 3714–3721.

43. Bui, M.H., D. Seligson, K.R. Han, *et al.* 2003. Carbonic anhydrase IX is an independent predictor of survival in advanced renal clear cell carcinoma: implications for prognosis and therapy. *Clin. Cancer Res.* **9:** 802–811.

44. Miyake, H., I. Sakai, M. Muramaki, *et al.* 2009. Prediction of response to combined immunotherapy with interferon-alpha and low-dose interleukin-2 in metastatic renal cell carcinoma: Expression patterns of potential molecular markers in radical nephrectomy specimens. *Int. J. Urol.* 2009 Apr 7.

45. Maruyama, R., K. Yamana, T. Itoi, *et al.* 2006. Absence of Bcl-2 and Fas/CD95/APO-1 predicts the response to immunotherapy in metastatic renal cell carcinoma. *Br. J. Cancer* **95:** 1244–1249.

46. Yap, T.A. & T.G. Eisen. 2006. Adjuvant therapy of renal cell carcinoma. *Clin. Genitourin. Cancer* **5:** 120–130.

47. Dorval, T., W.H. Fridman, C. Mathiot & P. Pouillart. 1992. Interleukin-2 therapy for metastatic uveal melanoma. *Eur. J. Cancer* **28A:** 2087.

48. Keilholz, U., C. Scheibenbogen, M. Brado, *et al.* 1994. Regional adoptive immunotherapy with interleukin-2 and lymphokine-activated killer (LAK) cells for liver metastases. *Eur. J. Cancer* **30A:** 103–105.

49. Phan, G.Q., P. Attia, S.M. Steinberg, *et al.* 2001. Factors associated with response to high-dose interleukin-2 in patients with metastatic melanoma. *J. Clin. Oncol.* **19:** 3477–3482.

50. Casamassima, A., M. Picciariello, M. Quaranta, *et al.* 2005. C-reactive protein: a biomarker of survival in patients with metastatic renal cell carcinoma treated with subcutaneous interleukin-2 based immunotherapy. *J. Urol.* **173:** 52–55.

51. Royal, R.E., S.M. Steinberg, R.S. Krouse, *et al.* 1996. Correlates of response to IL-2 therapy in patients treated for metastatic renal cancer and melanoma. *Cancer J. Sci. Am.* **2:** 91–98.

52. Sabatino, M., S. Kim-Schulze, M.C. Panelli, *et al.* 2009. Serum vascular endothelial growth factor and fibronectin predict clinical response to high-dose interleukin-2 therapy. *J. Clin. Oncol.* **27:** 2645–2652.

53. Tartour, E., J.Y. Blay, T. Dorval, *et al.* 1996. Predictors of clinical response to interleukin-2–based immunotherapy in melanoma patients: a French multiinstitutional study. *J. Clin. Oncol.* **14:** 1697–1703.

54. Bonfanti, A., P. Lissoni, R. Bucovec, *et al.* 2000. Changes in circulating dendritic cells and IL-12 in relation to the angiogenic factor VEGF during IL-2 immunotherapy of metastatic renal cell cancer. *Int. J. Biol. Markers* **15:** 161–164.

55. Cesana, G.C., G. DeRaffele, S. Cohen, *et al.* 2006. Characterization of CD4+CD25+ regulatory T cells in patients treated with high-dose interleukin-2 for metastatic melanoma or renal cell carcinoma. *J. Clin. Oncol.* **24:** 1169–1177.

56. Abrams, D. *et al.* 2009. Interleukin-2 therapy in patients with HIV infection. *N. Engl. J. Med.* **361**(16): 1548–1559.

# Perspective on Potential Clinical Applications of Recombinant Human Interleukin-7

**Claude Sportès,[a] Ronald E. Gress,[a] and Crystal L. Mackall[b]**

[a]*Experimental Transplantation and Immunology Branch, National Cancer Institute, National Institutes of Health, DHHS, Bethesda, Maryland, USA*

[b]*National Institutes of Health, DHHS, Bethesda, Maryland, USA*

**Interleukin-7 has critical and nonredundant roles in T cell development, hematopoiesis, and postdevelopmental immune functions as a prototypic homeostatic cytokine. Based on a large body of preclinical evidence, it may have multiple therapeutic applications in immunodeficiency states, either physiologic (immuno-senescence), pathologic (HIV) or iatrogenic (postchemotherapy and posthematopoietic stem cell transplant) and may have roles in immune reconstitution or enhancement of immunotherapy. Early clinical development trials in humans show that, within a short time, rhIL-7 administration results in a marked preferential expansion of both naive and memory CD4 and CD8 T cell pools with a tendency toward enhanced CD8 expansion. As a result, lymphopenic or normal older hosts develop an expanded circulating T cell pool with a profile that resembles that seen earlier in life with increased T cell repertoire diversity. These results, along with a favorable toxicity profile, open a wide perspective of potential future clinical applications.**

*Key words:* interleukin; cytokine; T cell

## Background

IL-7 is a multifunctional, homeostatic cytokine first isolated as a 25 KDa glycoprotein produced by a murine bone marrow stromal cell line.[1] IL-7 is not produced by lymphocytes but, rather, by bone marrow stroma[2] as well as other cell types including thymic stroma, keratinocytes, neurons, antigen presenting cells, lymph node follicular dendritic cells, and endothelial cells. IL-7 signals through a heterodimer involving IL-7Rα and the common γ chain (γc). IL-7Rα is shared with TSLP and the γc receptor is shared with IL-2, IL-4, IL-9, IL-15, and IL-21. IL-7 plays a critical, nonredundant role in the development of T cells. This is directly demonstrated in murine mod-

els wherein IL-7R knockout mice have arrest of T cell development at a double positive stage[3] and IL-7 deficient mice are profoundly lymphopenic with thymic cellularity reduced 20-fold.[4] In contrast, mice transgenic for IL-7 develop T cell lymphoproliferative/autoimmune diseases and T cell lymphomas.[5,6] The effects of IL-7 on T cell development are multifactorial, involving antiapoptotic and proliferative effects on developing lymphocytes,[7,8] promotion of V(D)J rearrangement of T cell receptor genes,[9] and provision of trophic and costimulatory signals for mature T cells.[10]

In human T cell development, IL-7's critical role is confirmed by analysis of three groups of patients with severe combined immunodeficiency (SCID) who have mutations involving the IL-7 receptor or its signaling pathway. Children with X-linked SCID (T-NK cell deficient, but spared B cells) have a defect of the γ-chain of the IL-7 receptor,[11] children with autosomal recessive SCID (T-NK cell deficient

Address for correspondence: Crystal L. Mackall, M.D., Chief, Pediatric Oncology Branch, National Cancer Institute, National Institutes of Health, DHHS, 10 Center Drive, CRC Room 1-3750, Bethesda, MD 20892-1104. Voice: 301 402 5940; fax: 301 451 7010. mackallc@mail.nih.gov

Cytokine Therapies: Ann. N.Y. Acad. Sci. 1182: 28–38 (2009).
doi: 10.1111/j.1749-6632.2009.05075.x © 2009 New York Academy of Sciences.

but spared B cells) not infrequently show mutations of the Jak-3 tyrosine kinase,[12,13] and kindreds with a defective α-chain of the IL-7 receptor are severely T cell deficient, but have normal NK cells and B cells.[14] Current concepts hold that the T cell deficiency in each of these clinical entities relates to absent or defective IL-7 signaling during T cell development, whereas NK cell deficiency results from absent or defective IL-15 signaling, which coexists when the genetic defect involves the γ-chain of the IL-7 receptor or the Jak-3 tyrosine kinase.

With regard to B cell development, IL-7 was first recognized as an important maturation and differentiation factor for pre-B cells in murine models[15] and IL-7 is an essential factor for supporting B lymphopoiesis *ex vivo*.[2] Moreover, IL-7 transgenic mice show expansion of immature B cells,[5,6] and humans treated with rhIL-7 show expansions in immature B cells within the bone marrow.[16] However, IL-7 does not appear to be essential for human B cell development, since patients with SCID due to γc, JAK3, or IL-7Rα mutations, can have normal or even elevated numbers of peripheral blood B cells.[11–14] Thus, while IL-7 participates in normal B cell development *ex vivo*,[17] it does not appear to be strictly required for B cell development in humans. IL-7 also stimulates egress of primitive hematopoietic cells from bone marrow, and it has been used successfully as a mobilization agent in mice, resulting in long lasting, full tri-lineage engraftment in mice transplanted with rhIL-7 mobilized peripheral blood, a property that may be clinically exploitable.[18]

In addition to its critical role in T cell lymphopoiesis and its effect on developing B cells, IL-7 also plays a central role in peripheral T cell homeostasis. As discussed above, IL-7R shares the receptor common γ-chain (CD132) with several other cytokines. Within this family, cytokines can be classified as activating versus homeostatic. IL-2 is a prototypic activating cytokine, while IL-7 is a prototypic homeostatic cytokine. IL-2 selectively signals activated T cells, is secreted by activated T cells and IL-2 signaling upregulates its own receptor (IL-2Rα; CD25) thus amplifying the IL-2 response during immune activation. In contrast, IL-7Rα (CD127) is expressed on resting T cells but is downregulated following signaling by IL-7 itself, other prosurvival cytokines (IL-2, IL-4, IL-6, IL-15)[19] or following T cell receptor (TCR) ligation. This tight regulation of IL-7Rα expression is congruent with the homeostatic role of IL-7, as it presumably prevents T cells that have already received a prosurvival signal from competing with other cells for its utilization.[20] Moreover, while production of activating cytokines occurs in the context of immune activation, IL-7 is continuously produced and available to resting T cells within the lymphoid niche. This continuous availability of IL-7 provides essential trophic signals for homeostatic proliferation and survival of naive T cells, since naive T cells adoptively transferred into an IL-7 deficient host rapidly disappear.[21]

Many studies have documented that IL-7 therapy can dramatically increase peripheral T cell numbers, primarily through augmentation of homeostatic peripheral expansion.[19,22,23] Briefly, homeostatic peripheral expansion is T cell receptor driven cycling, mediated primarily by low-affinity antigens. While augmentation of thymic output has also been recently suggested,[24] it remains controversial and very difficult to establish in the context of human trials. Regulatory CD4+CD25hi T cells (Tregs) express low IL-7Rα levels[25,26] and unlike IL-2, IL-7 therapy expands total CD4+ T cells without expanding Tregs.[16,27] IL-7 does not play a major role in Treg development, maintenance, and expansion; and in fact, recent evidence suggests that IL-7 may be capable of down modulating Treg activity.[28] In summary, in addition to potent effects on developing lymphocytes, IL-7 is required for maintenance of mature T cell populations and supraphysiologic levels of IL-7 expand peripheral T cell populations through a process termed homeostatic peripheral expansion.

## Potential Clinical Applications of IL-7 as an Immunorestorative

As individuals age, the adaptive immune system relies increasingly on the recruitment of memory cells to elicit immune responses. This is because the pool of naive T cells (i.e., the source of the wide diversity of specificities for antigen) decreases considerably with age. Furthermore, immune injury, whether physiologic (thymic involution with advancing age, immuno-senescence), pathologic (e.g. progressive immune depletion with HIV infection), or iatrogenic (following immune depleting therapy such as chemotherapy or irradiation), induces profound limitations in the natural pathways of T cell immune reconstitution. Immune reconstitution occurs through two primary pathways. The thymic pathway, which predominates in children, generates new T cells from pluripotent hematopoietic stem cells that home to the thymus and undergo expansion, differentiation, and selection. The resulting T cells, which bear a naive phenotype, display a diverse T cell receptor (TCR) repertoire and are poised to recognize an array of foreign antigens. In contrast, thymic-independent homeostatic peripheral expansion predominates in adults.[29] Homeostatic peripheral expansion results in a skewed T cell repertoire, which is poorly diversified and limited mostly to T cells that encounter their specific antigen during the period of immune reconstitution. Furthermore, homeostatic peripheral expansion is unable to restore numbers of CD4$^+$ T cells to pretreatment levels.[30]

Deficits in immune reconstitution are evident in multiple clinical settings wherein patients experience lymphocyte depletion. For instance, patients with human immunodeficiency virus infection, patients following allogeneic stem cell transplantation and older patients following high-dose cytotoxic therapy for cancer,[31–37] show incomplete or prolonged periods of lymphopenia before full immune reconstitution. Individuals older than 45 to 50 years of age, who experience lymphocyte depletion, are

likely to continue to have profound deficits in naïve T cells for the rest of their lives. While moderate immune competence can be accomplished through thymic-independent homeostatic peripheral expansion, new pathogens, or pathogens with high rates of mutations such as the influenza virus, may be a cause of substantial morbidity. Indeed, immune responses to immunizations such as influenza are diminished in elderly patients and subjects immunized after cancer chemotherapy appear to remain at increased risk of infection compared to controls (relative risk of developing protective titers ranges from 0.55 to 0.75, compared to a risk of 1 for normal control individuals).[38] Likewise, in HIV-infected individuals, antibody responses following immunization to T-dependent antigens are significantly decreased and correlate with the CD4 count.[39,40] Finally, cancer patients with limited immune reconstitution may also be at increased risk for tumor recurrence and are poor candidates for active immunotherapy strategies, which could potentially contribute to diminish disease recurrence.[35]

Because of the prevalence of long-term immune dysfunction in patients with age associated, iatrogenic or virus induced lymphopenia, there is great interest in administering immunorestoratives to hasten the capacity to restore normal immune function. RhIL-7 appears capable of substantially augmenting homeostatic peripheral expansion with preferential expansion of naive T cells, which bear the most diverse T cell receptor repertoires. Indeed, even athymic mice show fully restored immunocompetence with rhIL-7 therapy.[41] Thus, while it is not clear that rhIL-7 can reverse age, disease, or therapy-associated thymic involution, rhIL-7's capacity to augment naive cell proliferation may accomplish significant diversification of the T-cell receptor repertoire and restore near normal T cell diversity even in the absence of robust thymopoiesis.

Aging, in and of itself, even in the absence of lymphodepleting chemotherapy also poses a risk for diminished immune competence and preclinical data suggests that rhIL-7 therapy could be therapeutic in this setting. Age-associated abnormalities of both T cell and B cell compartments have been well described,[42,43] and humoral immune responses and immunoglobulin class switch are down-regulated in aged mice and humans.[44,45] There are many examples of a significant decrease in vaccine responses in the elderly population (tetanus and tick-borne encephalitis,[46] pneumococcal vaccine,[47] influenza).[48–53] Therefore, methods for improving vaccine efficiency in aged populations are critically needed. Given IL-7's potent capacity to augment responses to immunization and to augment repertoire diversity, it is plausible to consider IL-7 therapy as a means for augmenting vaccine responsiveness in elderly populations.

## Potential Role for IL-7 in Tumor-directe Immunotherapy

By enhancing immune reconstitution or expanding the immune cell repertoire, IL-7 could also have a significant role in enhancing immunotherapy for cancer.[19,22,23] IL-7 augments effector and memory responses to vaccination in mice[54] with preferential enhancement of responses to weak subdominant antigens, and improves survival of the CD8+ memory cell pool. In preclinical models, IL-7 therapy augments antitumor responses leading to improved survival when combined with antitumor vaccines.[54,55] When combined with tumor cell immunotherapy, IL-7 significantly prolongs the survival of tumor-bearing mice. This enhanced antitumor protection correlates with an increased number of activated dendritic cells and T cells in lymphoid tissues and increased activated effector T cells in the tumor microenvironment.[55]

The role of IL-7 in immunotherapy may go beyond the well-described enhancement of T cell numbers and T cell repertoire diversity, decreased apoptosis and increased sensitivity to CD3 trigger.[16] In a murine tumor model,[28]

IL-7 enhanced cytotoxic activity, increased the number of IL-17 producing CD4+ T cells, and increased serum cytokine levels (IL-6, IL-1α, IL-1β, IL-12, tumor necrosis factor-α, C-C chemokine ligand-5 (RANTES), macrophage inflammatory protein-1α). Moreover, IL-7 appeared to inhibit Treg function, and increased refractoriness to inhibitory signals by abrogating TGF-β induced inhibition of CD8+ T cell proliferation and mediating anti-TGF-β effects through down modulation of Cbl-b expression. Finally, previous work has demonstrated that IL7-Rα+ expression on activated T cells is associated with increased central memory cell generation[56] and IL7-Rα+ expression on adoptively transferred antigen-specific T cells correlates with better survival *in vivo*.[57,58] Thus, it is possible that rhIL-7 therapy could augment the effectiveness of adoptive T cell therapy for cancer by improving survival of IL-7-Rα+ central memory populations and diminishing the competitiveness of senescent IL-7-Rα− populations. Finally, recent studies have suggested that rhIL-7 can diminish PD-1 expression on activated CD8+ populations,[28] an effect which would be predicted to enhance CD8+ T cell survival *in vivo*.

## RhIL-7 in Clinical Trials

The first five studies of rhIL-7 initiated in humans evaluated an E. Coli produced, non-glycosylated rhIL-7 ("CYT99 007," Cytheris Inc., Rockville, MD) using various phase I designs with doses ranging from 3 to 60 μg/kg in three different populations and dose schedules (see Table 1). Two trials involved oncology subjects, two trials involved HIV+ subjects, and one trial involved subjects following allogeneic transplantation for nonlymphoid malignancy. In Oncology trial 1, careful immunologic studies were performed on all IL-7 recipients allowing detailed analysis of the effects of rhIL-7 therapy in humans. All subjects treated between 10–60 mcg/kg/dose showed dose dependent increases in circulating

**TABLE 1.** Summary of Clinical Studies with rhIL-7 "CYT 99 007"

| TRIAL | "Oncology 1" | "Oncology 2" | "HIV 1" | "HIV 2" | "Allogeneic transplantation" |
|---|---|---|---|---|---|
| Institution | NCI; ETIB, POB | NCI; SB | Case Western; Cleveland, OH, NIAID; Bethesda, MD | Hôpital Henri-Mondor, Creteil, France | MSKCC |
| | Bethesda, MD | Bethesda, MD | | | New York, NY |
| Investigator | Sportès, C. | Rosenberg, S. | Lederman, M., Sereti, I. | Lévy, Y. | van den Brink, M. |
| Subjects | Refractory malignancy CD3 > 300/mm$^3$ | Refractory metastatic melanoma | HIV: HAART > 12 mo. CD4: > 100/mm$^3$ | HIV : HAART > 12 mo. CD4: 100 – 400/mm$^3$ | Myeloid leukemia post allogeneic transplantation |
| Number of subjects | 16 | 12 | 17 (stratum1) 8 (stratum 2) | 14 | 1 |
| Design | Phase I: dose escalation | Phase I: dose escalation (co-administration of gp 100/MART peptide vaccines) | IL-7/placebo (random 3 to 1) 2 strata: HIV RNA: • < 50 or • 50–50,000 copies | Phase I: dose escalation 2 strata: CD4/mm$^3$ • 100–200 • 200–400 | Phase I: dose escalation |
| Doses/ schedule | • 3, 10, 30, 60 μg/kg/dose • every other day; • 8 doses (2 weeks) | • 3, 10, 30, 60 μg/kg/dose • 3 times a week • 8 doses (3 weeks) | • 3, 10, 30, 60 μg/kg/dose • single dose | • 3, 10 μg/kg/dose • 3 times a week • 8 doses (3 weeks) | • 3 μg/kg/dose • 3 times a week • 8 doses (3 weeks) |
| Reference | [16] | [27] | [60] | [59] | |

NCI: National Cancer Institute; ETIB: Experimental Transplantation and Immunology Branch; POB: Pediatric Oncology Branch; SB: Surgery Branch; NIAID: National Institute of Allergy and Infectious Diseases; MSKCC: Memorial Sloan-Kettering Cancer Center.

absolute lymphocyte counts, and lymphoid organ enlargement (spleen and normal lymph nodes, but not thymus) was seen at the doses of 30 and 60 μg/kg/dose, maximum on day 14, then returning to baseline over several weeks (Fig. 1). In most subjects, CD3$^+$ αβ and γδ cells, CD4$^+$ and CD8$^+$ T cells increased equally, in a clear dose-dependent fashion. The absolute cell numbers peaked one week after the end of treatment (day 21) but remained elevated for up to 2 months following the end of treatment. Similarly, increased lymphocyte counts persisted for 24 and 48 weeks after the last administered rhIL-7 dose in HIV individuals with the longest follow-up.[59] There was no correlation between subject age and magnitude of the increase in CD3$^+$, CD4$^+$, or CD8$^+$ T cells.

The effects of rhIL-7 on T cell expansion could be attributed to a combination of increased cell cycling and diminished programmed cell death but were self-limited by downregulation of IL-7Rα. At baseline, 1%–3% of circulating CD4$^+$ and CD8$^+$ T cells were in cycle (expressing Ki-67). Therapy resulted in a dramatic, dose-dependent increase in cycling frequency: in subjects treated with 60 μg/kg/dose, >40% of CD4$^+$ peripheral T cells and >55% of CD8$^+$ peripheral T cells expressed Ki-67 at day 7. T cell cycling substantially declined by day 14, coincident with maximum IL-7Rα downregulation despite sustained pharmacologic serum IL-7 levels (continued rhIL-7 administration through day 14). As a surrogate for decreased apoptosis, bcl-2 expression was increased and

**(A)** **Peripheral blood cell number per mm³:** *percent change over baseline*

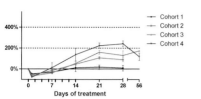
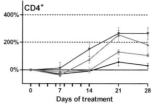
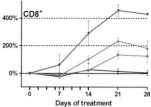

**(B)** **Spleen Size**
*(percent change over baseline)*

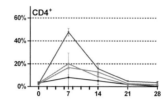

**(C)** **Ki 67** *(percent of subset expressing Ki67)*

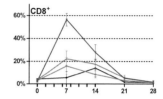

**(D)** **IL-7Rα** *(percent of subset expressing CD127)*

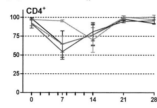
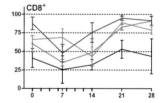

**(E)** **Bcl-2** *(mean fluorescence intensity)*

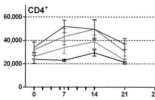
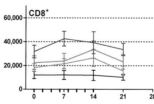

**Figure 1.** Effects of rhIL-7 Therapy on Circulating lymphocytes and Spleen Size. RhIL-7 injections indicated by tick marks on X-axis along with baseline, days 7, 14, 21, and 28 data points (additional data points are shown for total lymphocytes count only: day 1 for all and day 55–90 for subjects treated with 30 or 60 μg/kg/d). Mean value for each cohort (± SEM) are plotted: 3 μg/kg/d (black); 10 μg/kg/d (green); 30 μg/kg/d (red); 60 μg/kg/d (blue). **(A)** Absolute lymphocyte count from complete blood counts (*left panels*) and percent change in absolute numbers over baseline (*right panels*) for CD3⁺/CD4⁺ and CD3⁺/CD8⁺ cells the respective subsets are shown. **(B)** Spleen size: percent changes from the pretherapy bidimensional product by CT scan. **(C, D,** and **E)** CD3⁺/CD4⁺ (*left panels*) and CD3⁺/CD8⁺ subsets (*right panels*). **(C)** "Ki-67": percentage of cells expressing Ki-67 (flow cytometry). **(D)** "IL-7Rα" expression as mean fluorescence intensity via flow cytometry. **(E)** "bcl-2" expression as mean fluorescence intensity (after subtracting background staining for each subset) via flow cytometry.

remained elevated throughout the rhIL-7 administration, likely contributing to the increase in T cell numbers. Indeed, at the time of peak lymphocyte expansion, one week after the end of the treatment (day 21), cell cycling had returned to baseline values.[16]

Detailed studies of T cell subsets[16] showed that rhIL-7 preferentially expanded

(A) **cell number**

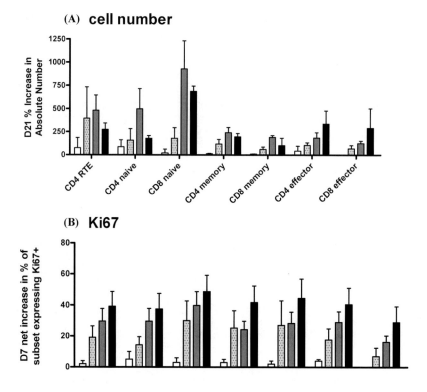

(B) **Ki67**

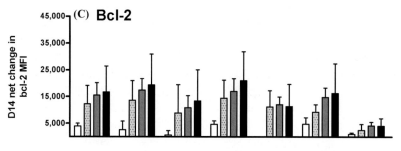

(C) **Bcl-2**

(D) **cell number** (subsets defined by CCR7/CD45RA staining)

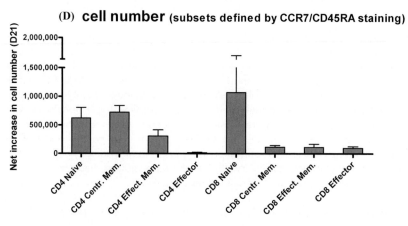

the CD4$^+$ recent thymic emigrants (CD4$^+$/CD31$^+$/CD45RA$^+$), naive (CD4$^+$/CD27$^+$/CD45RA$^+$), and central memory (CD4$^+$/CD45RA$^-$/CCR7$^+$) T cells as well as CD8$^+$ naive (CD8$^+$/CD27$^+$/CD45RA$^+$) T cells (Fig. 2). This T cell expansion resulted in a statistically significant increase in T cell repertoire diversity for both CD4$^+$ and CD8$^+$ T cells along with a decrease in the proportion of terminally differentiated effector T cells. It is also noteworthy that rhIL-7 preferentially expanded naive and central memory subsets, both of which are important in initiating lymph node germinal center formation. Tregs showed markedly less proliferation and expansion compared to other CD4$^+$ T cells, resulting in a decrease in the proportion of circulating regulatory T cells. Therefore, within the short time frame of the study, rhIL-7 therapy induced marked expansion of circulating T cells exhibiting a rejuvenated profile resembling that seen early in life with minimal Treg expansion. These lymphocytes remained functional with conserved or increased *in vitro* responsiveness to anti-CD3 stimulation. Ongoing studies are underway to determine whether combining antiviral therapy with rhIL-7 may augment viral clearance in hepatitis C infection and whether rhIL-7 therapy can safely augment immune reconstitution following T cell depleted allogeneic SCT.

primates. IL-7 potently expands T cell populations, predominantly through enhanced peripheral T cell cycling. As a result, lymphopenic hosts show normalization of T cell numbers with rhIL-7 therapy, and normal hosts show supranormal T cell numbers with rhIL-7. In general, both CD4 and CD8 populations are expanded *in vivo*, with a tendency toward enhanced CD8 expansion. Both naive and memory subsets undergo expansion, but there appears to be a preferential expansion of naive subsets and little if any expansion of senescent memory populations. T cell receptor repertoire diversity can be enhanced by rhIL-7 therapy, even in the absence of direct thymopoietic effects. Tregs are not targeted by rhIL-7, leading to a relative reduction in Treg frequency when rhIL-7 therapy is administered. RhIL-7 shows promise as an immunorestorative for aged individuals or individuals who have experienced iatrogenic or disease induced lymphocyte depletion. It also is active as a vaccine adjuvant in preclinical models, and clinical trials are underway to determine whether it may augment viral clearance in the context of chronic viral infection when used in conjunction with antiviral therapy. Future studies will seek to determine whether rhIL-7 may enhance the effectiveness of tumor vaccines and/or adoptive immunotherapy for cancer.

## Summary

RhIL-7 has now entered clinical trials and shows clear evidence for biologic activity that largely mirrors the results founds in mice and

### Conflicts of Interest

The authors declare no conflicts of interest.

**Figure 2.** RhIL-7 induces preferential expansion of Naive T cells and CD4$^+$ Memory T Cells (with cycling and bcl-2 expression increase across subsets).[16] Subjects treated at 3 µg/kg/d (white), 10 µg/kg/d (speckled), 30 µg/kg/d (grey) and 60 µg/kg/d (black). The time points shown represent the point of maximal increase for each parameter: (**A**) Percent increase in absolute circulating cell number (/mm$^3$) over pretherapy value for each subset at Day 21 following rhIL-7. (**B**) Net increase from baseline in the proportion of Ki67$^+$ cells in each subset at day 7 of rhIL-7 therapy. (**C**) Net increase in bcl-2 mean fluorescence intensity in each subset at day 14. (**D**) Net increase in cell number at day 21 in subsets defined by CCR7 and CD45RA surface expression as follows: Naive (CCR7$^+$CD45RA$^+$), central memory (CCR7$^+$CD45RA$^-$), effector memory (CCR7$^-$CD45RA$^-$), and effector RA (CCR7$^-$CD45RA$^+$).

# References

1. Goodwin, R.G., S. Lupton, A. Schmierer, *et al.* 1989. Human interleukin 7: molecular cloning and growth factor activity on human and murine B-lineage cells. *Proc. Natl. Acad. Sci. USA* **86:** 302–306.

2. Sudo, T., M. Ito, Y. Ogawa, *et al.* 1989. Interleukin 7 production and function in stromal cell-dependent B cell development. *J. Exp. Med.* **170:** 333–338.

3. Peschon, J.J., P.J. Morrissey, K.H. Grabstein, *et al.* 1994. Early lymphocyte expansion is severely impaired in interleukin 7 receptor-deficient mice. *J. Exp. Med.* **180:** 1955–1960.

4. Freeden-Jeffry, U., P. Vieira, L.A. Lucian, *et al.* 1995. Lymphopenia in interleukin (IL)-7 gene-deleted mice identifies IL-7 as a nonredundant cytokine. *J. Exp. Med.* **181:** 1519–1526.

5. Fisher, A.G., C. Burdet, C. Bunce, *et al.* 1995. Lymphoproliferative disorders in IL-7 transgenic mice: expansion of immature B cells which retain macrophage potential. *Int. Immunol.* **7:** 415–423.

6. Samaridis, J., G. Casorati, A. Traunecker, *et al.* 1991. Development of lymphocytes in interleukin 7-transgenic mice. *Eur. J. Immunol.* **21:** 453–460.

7. Maraskovsky, E., L.A. O'Reilly, M. Teepe, *et al.* 1997. Bcl-2 can rescue T lymphocyte development in interleukin-7 receptor- deficient mice but not in mutant rag-1-/- mice. *Cell* **89:** 1011–1019.

8. Akashi, K., M. Kondo, U. Freeden-Jeffry, *et al.* 1997. Bcl-2 rescues T lymphopoiesis in interleukin-7 receptor-deficient mice. *Cell* **89:** 1033–1041.

9. Muegge, K., M.P. Vila & S.K. Durum. 1993. Interleukin-7: A cofactor for V(DJ) rearrangement of the T-cell receptor á gene. *Science* **261:** 93–95.

10. Yeoman, H., D.R. Clark & D. DeLuca. 1996. Development of CD4 and CD8 single positive T cells in human thymus organ culture: IL-7 promotes human T cell production by supporting immature T cells. *Dev. Comp. Immunol.* **20:** 241–263.

11. Noguchi, M., H. Yi, H.M. Rosenblatt, *et al.* 1993. Interleukin-2 receptor gamma chain mutation results in X-linked severe combined immunodeficiency in humans. *Cell* **73:** 147–157.

12. Macchi, P., A. Villa, S. Giliani, *et al.* 1995. Mutations of Jak-3 gene in patients with autosomal severe combined immune deficiency (SCID). *Nature* **377:** 65–68.

13. Russell, S.M., N. Tayebi, H. Nakajima, *et al.* 1995. Mutation of Jak3 in a patient with SCID: essential role of Jak3 in lymphoid development. *Science* **270:** 797–800.

14. Puel, A., S.F. Ziegler, R.H. Buckley, *et al.* 1998. Defective IL7R expression in T(-)B(+)NK(+) severe combined immunodeficiency. *Nat. Genet.* **20:** 394–397.

15. Sudo, T., S. Nishikawa, N. Ohno, *et al.* 1993. Expression and function of the interleukin 7 receptor in murine lymphocytes. *Proc. Natl. Acad. Sci. USA* **90:** 9125–9129.

16. Sportes, C., F.T. Hakim, S.A. Memon, *et al.* 2008. Administration of rhIL-7 in humans increases in vivo TCR repertoire diversity by preferential expansion of naive T cell subsets. *J. Exp. Med.* **205:** 1701–1714.

17. Dittel, B.N. & T.W. LeBien. 1995. The growth response to IL-7 during normal human B cell ontogeny is restricted to B-lineage cells expressing CD34. *J. Immunol.* **154:** 58–67.

18. Grzegorzewski, K.J., K.L. Komschlies, S.E. Jacobsen, *et al.* 1995. Mobilization of long-term reconstituting hematopoietic stem cells in mice by recombinant human interleukin 7. *J. Exp. Med.* **181:** 369–374.

19. Fry, T.J., M. Moniuszko, S. Creekmore, *et al.* 2003. IL-7 therapy dramatically alters peripheral T-cell homeostasis in normal and SIV-infected nonhuman primates. *Blood* **101:** 2294–2299.

20. Park, J.H., Q. Yu, B. Erman, *et al.* 2004. Suppression of IL7Ralpha transcription by IL-7 and other prosurvival cytokines: a novel mechanism for maximizing IL-7-dependent T cell survival. *Immunity* **21:** 289–302.

21. Tan, J.T., E. Dudl, E. LeRoy, *et al.* 2001. IL-7 is critical for homeostatic proliferation and survival of naive T cells. *Proc. Natl. Acad. Sci. USA* **98:** 8732–8737.

22. Chu, Y.W., S.A. Memon, S.O. Sharrow, *et al.* 2004. Exogenous IL-7 increases recent thymic emigrants in peripheral lymphoid tissue without enhanced thymic function. *Blood* **104:** 1110–1119.

23. Storek, J., T. Gillespy, III, H. Lu, *et al.* 2003. Interleukin-7 improves CD4 T-cell reconstitution after autologous CD34 cell transplantation in monkeys. *Blood* **101:** 4209–4218.

24. Beq, S., M.T. Nugeyre, F.R. Ho Tsong, *et al.* 2006. IL-7 induces immunological improvement in SIV-infected rhesus macaques under antiviral therapy. *J. Immunol.* **176:** 914–922.

25. Liu, W., A.L. Putnam, Z. Xu-Yu, *et al.* 2006. CD127 expression inversely correlates with FoxP3 and suppressive function of human CD4+ T reg cells. *J. Exp. Med.* **203:** 1701–1711.

26. Seddiki, N., B. Santner-Nanan, J. Martinson, *et al.* 2006. Expression of interleukin (IL)-2 and IL-7 receptors discriminates between human regulatory and activated T cells. *J. Exp. Med.* **203:** 1693–1700.

27. Rosenberg, S.A., C. Sportes, M. Ahmadzadeh, *et al.* 2006. IL-7 administration to humans leads to expansion of CD8+ and CD4+ cells but a relative decrease of CD4+ T-regulatory cells. *J. Immunother.* **29:** 313–319.

28. Pellegrini, M., T. Calzascia, A.R. Elford, *et al.* 2009. Adjuvant IL-7 antagonizes multiple cellular and molecular inhibitory networks to enhance immunotherapies. *Nat. Med.* **15:** 528–536.

29. Mackall, C.L., L. Granger, M.A. Sheard, *et al*. 1993. T-cell regeneration after bone marrow transplantation: differential CD45 isoform expression on thymic-derived versus thymic-independent progeny. *Blood* **82:** 2585–2594.

30. Mackall, C.L., C.V. Bare, L.A. Granger, *et al*. 1996. Thymic-independent T cell regeneration occurs via antigen-driven expansion of peripheral T cells resulting in a repertoire that is limited in diversity and prone to skewing. *J. Immunol.* **156:** 4609–4616.

31. Mackall, C.L., T.A. Fleisher, M.R. Brown, *et al*. 1994. Lymphocyte depletion during treatment with intensive chemotherapy for cancer. *Blood* **84:** 2221–2228.

32. Hakim, F.T., R. Cepeda, S. Kaimei, *et al*. 1997. Constraints on CD4 recovery postchemotherapy in adults: thymic insufficiency and apoptotic decline of expanded peripheral CD4 cells. *Blood* **90:** 3789–3798.

33. Mackall, C.L., T.A. Fleisher, M.R. Brown, *et al*. 1997. Distinctions between CD8+ and CD4+ T-cell regenerative pathways result in prolonged T-cell subset imbalance after intensive chemotherapy. *Blood* **89:** 3700–3707.

34. Mackall, C.L. 2000. T-cell immunodeficiency following cytotoxic antineoplastic therapy: a review. *Stem Cells* **18:** 10–18.

35. Hakim, F.T. & R.E. Gress. 2005. Reconstitution of the lymphocyte compartment after lymphocyte depletion: a key issue in clinical immunology. *Eur. J. Immunol.* **35:** 3099–3102.

36. Hakim, F.T., S.A. Memon, R. Cepeda, *et al*. 2005. Age-dependent incidence, time course, and consequences of thymic renewal in adults. *J. Clin. Invest.* **115:** 930–939.

37. Sportes, C., N.J. McCarthy, F. Hakim, *et al*. 2005. Establishing a platform for immunotherapy: clinical outcome and study of immune reconstitution after high-dose chemotherapy with progenitor cell support in breast cancer patients. *Biol. Blood Marrow Transplant.* **11:** 472–483.

38. Gross, P.A., A.L. Gould & A.E. Brown. 1985. Effect of cancer chemotherapy on the immune response to influenza virus vaccine: review of published studies. *Rev. Infect. Dis.* **7:** 613–618.

39. Kroon, F.P., J.T. van Dissel, J.C. de Jong, *et al*. 1994. Antibody response to influenza, tetanus and pneumococcal vaccines in HIV-seropositive individuals in relation to the number of CD4+ lymphocytes. *AIDS (London, England)* **8:** 469–476.

40. Malaspina, A., S. Moir, S.M. Orsega, *et al*. 2005. Compromised B cell responses to influenza vaccination in HIV-infected individuals. *J. Infect. Dis.* **191:** 1442–1450.

41. Fry, T.J., B.L. Christensen, K.L. Komschlies, *et al*. 2001. Interleukin-7 restores immunity in athymic T-cell-depleted hosts. *Blood* **97:** 1525–1533.

42. Johnson, S.A. & J.C. Cambier. 2004. Ageing, autoimmunity and arthritis: senescence of the B cell compartment – implications for humoral immunity. *Arthritis Res. Ther.* **6:** 131–139.

43. Fulop, T., A. Larbi, A. Wikby, *et al*. 2005. Dysregulation of T-cell function in the elderly : scientific basis and clinical implications. *Drugs & Aging.* **22:** 589–603.

44. Frasca, D., D. Nguyen, R.L. Riley, *et al*. 2003. Decreased E12 and/or E47 transcription factor activity in the bone marrow as well as in the spleen of aged mice. *J. Immunol.* **170:** 719–726.

45. Frasca, D., R.L. Riley & B.B. Blomberg. 2005. Humoral immune response and B-cell functions including immunoglobulin class switch are downregulated in aged mice and humans. *Semin. Immunol.* **17:** 378–384.

46. Hainz, U., B. Jenewein, E. Asch, *et al*. 2005. Insufficient protection for healthy elderly adults by tetanus and TBE vaccines. *Vaccine* **23:** 3232–3235.

47. Artz, A.S., W.B. Ershler & D.L. Longo. 2003. Pneumococcal vaccination and revaccination of older adults. *Clin. Microbiol. Rev.* **16:** 308–318.

48. Hirota, Y., M. Kaji, S. Ide, *et al*. 1996. The hemagglutination inhibition antibody responses to an inactivated influenza vaccine among healthy adults: with special reference to the prevaccination antibody and its interaction with age. *Vaccine* **14:** 1597–1602.

49. Gross, P.A., C. Russo, M. Teplitzky, *et al*. 1996. Time to peak serum antibody response to influenza vaccine in the elderly. *Clin. Diagnostic Lab. Immunol.* **3:** 361–362.

50. Murasko, D.M., E.D. Bernstein, E.M. Gardner, *et al*. 2002. Role of humoral and cell-mediated immunity in protection from influenza disease after immunization of healthy elderly. *Exp. Gerontol.* **37:** 427–439.

51. Fulop, T. Jr., J.R. Wagner, A. Khalil, *et al*. 1999. Relationship between the response to influenza vaccination and the nutritional status in institutionalized elderly subjects. *J. Gerontol. A Biol. Sci. Med. Sci.* **54:** M59–M64.

52. Nichol, K.L., J.D. Nordin, D.B. Nelson, *et al*. 2007. Effectiveness of influenza vaccine in the community-dwelling elderly. *N. Engl. J. Med.* **357:** 1373–1381.

53. Simonsen, L., R.J. Taylor, C. Viboud, *et al*. 2007. Mortality benefits of influenza vaccination in elderly people: an ongoing controversy. *Lancet Infect Dis.* **7:** 658–666.

54. Melchionda, F., T.J. Fry, M.J. Milliron, *et al*. 2005. Adjuvant IL-7 or IL-15 overcomes immunodominance and improves survival of the CD8+ memory cell pool. *J. Clin. Invest.* **115:** 1177–1187.

55. Li, B., M.J. Vanroey & K. Jooss. 2007. Recombinant IL-7 enhances the potency of GM-CSF-secreting

tumor cell immunotherapy. *Clin. Immunol.* **123:** 155–165.

56. Kaech, S.M., J.T. Tan, E.J. Wherry, *et al.* 2003. Selective expression of the interleukin 7 receptor identifies effector CD8 T cells that give rise to long-lived memory cells. *Nat. Immunol.* **4:** 1191–1198.

57. Gattinoni, L., S.E. Finkelstein, C.A. Klebanoff, *et al.* 2005. Removal of homeostatic cytokine sinks by lymphodepletion enhances the efficacy of adoptively transferred tumor-specific CD8$^+$ T cells. *J. Exp. Med.* **202:** 907–912.

58. Powell, D.J., Jr., M.E. Dudley, P.F. Robbins, *et al.* 2005. Transition of late-stage effector T cells to CD27$^+$ CD28$^+$ tumor-reactive effector memory T cells in humans after adoptive cell transfer therapy. *Blood* **105:** 241–250.

59. Levy, Y., C. Lacabaratz, L. Weiss, *et al.* 2009. Enhanced T cell recovery in HIV-1-infected adults through IL-7 treatment. *J. Clin. Invest.* **119:** 997–1007.

60. Sereti, I., R.M. Dunham, J. Spritzler, *et al.* 2009. IL-7 administration drives T cell cycle entry and expansion in HIV-1 infection. *Blood*

# Immune Modulation with Interleukin-21

## Neela S. Bhave and William E. Carson III

*The Ohio State University, Comprehensive Cancer Center, Columbus, Ohio, USA*

Interleukin 21 (IL-21) is produced by activated CD4[+] T cells. The IL-21R shares the common receptor gamma-chain with IL-2, IL-4, IL-7, IL-9, and IL-15, is widely expressed on immune cells, and mediates a variety of effects on the immune system. IL-21 enhances the proliferation, antigen-induced activation, clonal expansion, IFN-γ production, and cytotoxicity of NK cells and T cells. The antitumor actions of IL-21 have been variously attributed to NK cell and CD8[+] T cell cytotoxicity, CD4[+] T cell help, NKT cells, and the antiangiogenic properties induced by IFN-γ secretion. In clinical trials IL-21 has been well tolerated and induces a unique pattern of immune activation. IL-21 is therefore an excellent candidate for use in immune therapy.

*Key words:* interleukin-21; tumor immunology; innate immunity; NK cells

## Introduction

Interleukin-21 (IL-21) is produced primarily by activated CD4[+] T cells. The IL-21 receptor (IL-21R) shares the common receptor gamma-chain with IL-2, IL-4, IL-7, IL-9, and IL-15. The IL-21R is widely expressed on immune cells and mediates a variety of effects depending on the cell type under study. The IL-21R is a type I cytokine receptor with four conserved cysteine residues and an extracellular WSXWS motif.[1] It is most closely related to IL-2Rβ and IL-4Rα.[1,2] An important property of IL-21 is its ability to enhance the proliferation, antigen-induced activation, clonal expansion, interferon-gamma (IFN-γ) production, and cytotoxicity of CD8[+] T cells. In addition to its effects on CD4[+] and CD8[+] T cells, IL-21 has context-dependent effects on B cells (such as the ability to induce granzyme B), amplifies macrophage activation pathways, and inhibits the activation of myeloid dendritic cells (DCs).[1,3–5]

IL-21 promotes the maturation of natural killer (NK) cells and downregulates the expression of NKG2D on human NK cells while increasing the expression of the NK activation receptors NKp30 and 2B4.[6,7] Co-stimulation of NK cells with IL-21 and IL-15 and/or IL-18 is also an effective stimulus for IFN-γ production.[8]

IL-21 has been implicated in several disease processes ranging from allergy, autoimmune disorders, and viral infections to cancer. Importantly, IL-21 is able to mediate the regression of established tumors in a variety of murine models. Depending on the method of cytokine delivery (exogenous cytokine, plasmid DNA, or retroviral transduction of tumor cells) its mechanism of action has been variously attributed to NK cell cytolytic activity, perforin-mediated CD8[+] T cell cytotoxicity, CD4[+] T cell help, NKT cells and the antiangiogenic actions of IL-21-induced IFN-γ.[9–14] Phase I trials have revealed that IL-21 is well tolerated and has clinical activity as a single agent.[15] Clinical trials are underway to assess clinical safety of IL-21 as a single agent and when used in combination with therapeutic monoclonal antibodies and other treatments (Clinical trial identifier number NCT00347971).

Address for correspondence: William E. Carson III, OSU Comprehensive Cancer Center, The Ohio State University, N924 Doan Hall, 410 W. 10th Ave., Columbus, OH 43210. Voice: 614-292-5819; fax: 614-688-4366. william.carson@osumc.edu

Cytokine Therapies: Ann. N.Y. Acad. Sci. 1182: 39–46 (2009).
doi: 10.1111/j.1749-6632.2009.05071.x © 2009 New York Academy of Sciences.

## IL-21 Signal Transduction

IL-21R is a heterodimer of the IL-21R $\alpha$-chain and the common $\gamma$-chain. The signaling component of the IL-21R complex is the common $\gamma$-chain, which is also a functional component of the IL-2R, IL-4R, and IL-15R.[16] The IL-21R$\alpha$ contains six tyrosine residues in its cytoplasmic domain.[4] The binding of IL-21 to the IL-21R results in the autophosporylation of Janus-activated kinase1 (Jak1) and Jak3. The autophosphorylation of Jak1 and Jak3 leads to activation and nuclear translocation of signal transducer and activator of transcription 1 (STAT1), STAT3, and STAT5 proteins and transcription of IL-21-responsive genes.[5] IL-21-induced expression of IFN-$\gamma$ in NK cells and T cells has been shown to be dependent upon the activation of STAT proteins.[8] IL-21R induces the expression of suppressor of cytokine signaling1 (*SOCS1*) in CD8$^+$ T cells, an event which could be due to STAT3 phosphorylation.[17,18] SOCS1 deactivates phosphorylated Jak-STAT signaling intermediates and therefore acts as negative regulator of IL-21-induced signal transduction.[8,19]

## Regulation of Specific Immunity by IL-21

### B Cells

The immune effects of IL-21 depend on the type, activation status, and stage of differentiation of the cells under study. IL-21 plays an important role in the development and differentiation of B cells.[20] In *in vitro* experiments, IL-21 enhanced the anti-CD40 mediated proliferation of human B cells but inhibited IL-4 and anti-IgM antibody mediated proliferation.[1,21,22] Less consistent effects were observed in murine B cells.[21] Of note, IL-21 induced the apoptosis of resting primary murine B cells. IL-21-mediated apoptosis was not affected by prestimulation of primary B cells with IL-4, LPS, or anti-CD40 antibody. The induc-

tion of apoptosis by IL-21 correlated with a downregulation in the expression of two anti-apoptotic genes of the Bcl-2 family Bcl-2 and Bcl-x(L).[23] Overall, IL-21 seems to have an immunosuppressive effect on B cells.

### T Cells

IL-21 appears to be a key factor for maturation of the adaptive T cell immune responses. IL-21 is produced by CD4$^+$ T cells, specially Th1, Th2, and Th17 T cell subsets as well as NKT cells.[1,24–27] In general, IL-21 may be described as a T-cell co-stimulatory molecule. The IL-21R is present on both CD4$^+$ and CD8$^+$ T cells.[28] Treatment of naïve CD4$^+$ T cells with IL-21 downregulates IFN-$\gamma$ production without affecting the production of other Th1 cytokines.[29] Stimulation of CD8$^+$ T cells with IFN-$\alpha$ and IL-21 triggers increased STAT3 activation. This costimulation also triggers a selective increase in MHC class I expression and NK and CD8$^+$ T-cell-mediated cytotoxicity.[30] Ma *et al.*[10] used a cytokine gene therapy approach to study the antitumor responses mediated by IL-21 in the B16F1 melanoma and MethA fibrosarcoma tumor models in mice. Tumors were retrovirally transduced to secrete IL-21. Rejection of IL-21-secreting tumors required the presence of the IL-21R and was not dependent on helper CD4$^+$ T cells. Thedrez and colleagues subsequently showed that IL-21 played an important role in the cytotoxicity and Th1 programming of precommitted antigen-stimulated $\gamma\delta$T cells and thereby enhanced $\gamma\delta$T cell-mediated antitumor responses.[31] IL-21 might be important in maintaining the balance between Th1 and Th2 T cell activity which can be beneficial in clinic.

IL-21 plays an important role in the regulation of Th17 development. Th17 commitment is regulated by IL-21-induced upregulation of the IL-23R. IL-23 is important in the development of Th17 cells, but its receptor is not expressed on naïve T cells.[32] Therefore, IL-21 mediated induction of IL-23R may be critical for Th17 development and differentiation. IL-21 is also involved in regulating the development of another subset of T cells known as regulatory T cells (Tregs). Tregs are CD4$^+$ CD25$^+$ subset of T cell that suppresses the immune response of other cells. T cell receptor (TCR) activation stimulates the upregulation of IL-21R on naïve Treg cells. IL-2, IL-7, and IL-15 induce proliferation of Treg cells, but IL-21 actually works to inhibit the suppressive function of Treg cells.[33,34] IL-21 can overcome Treg mediated immunosuppression.

## Role of IL-21 in Innate Immunity

The innate immune system is made up of cells that can respond to microbial challenges without the need for prior sensitization. The cells of the innate immune system, for example, NK cells, monocytes, macrophages, and dendritic cells participate in an immediate, non-specific response to microbial invaders and also promote the development of specific immune response.[35] IL-21 has pleotropic effects on multiple cellular components of the innate immune response.

## Dendritic Cells (DCs)

DCs are involved in antigen presentation to cells of the adaptive immune system.[36] DCs induce antigen-specific T cell responses. IL-21 strongly induces expression of SOCS genes in monocyte-derived DCs. *SOCS1* and *SOCS3* negatively regulate TLR signaling. Specifically, lipopolysaccharide (LPS)-induced tumor necrosis factor-$\alpha$ production is inhibited by IL-21 in a SOCS-dependent manner. IL-21 also inhibits LPS-induced secretion of IL-12, CCL5 and CXCL10 by monocyte-derived DCs.[18] DCs generated with granulocyte macrophage-colony stimulating factor (GM-CSF) in the presence of IL-21 (IL-21DCs) had reduced major histocompatibility complex class II (MHC II) expression, high antigen uptake, and low stimulatory capacity for T-cell activation *in vitro*. IL-21DCs completely failed to induce antigen-specific T-cell mediated contact hypersensitivity. IL-21 negatively regulates DC-T cell interaction.[37] Thus it appears that IL-21 inhibits DC activation,[38] making it a potential tool for therapeutic manipulation of DC-induced immune reactions.[39]

## Natural Killer (NK) Cells

NK cells do not express T-cell receptors (TCR), but they do mediate perforin-dependent cell lysis depending on the engagement of their activating or inhibitory receptors by target cell class I molecules. Numerous innate immunity functions are attributed to NK cells. Among these are the stimulation of innate phagocytes via IFN-$\gamma$ secretion and the destruction of virally infected cells via granzyme and perforin release. Their cytokine production also appears to be important in priming the antigen-specific T cell response. The secretion of these inflammatory cytokines from NK cells dramatically increases the cytotoxic response against tumor cells and leads to their killing.

NK cells express a low affinity activating receptor Fc$\gamma$RIIIa (CD16) which binds to the Fc region of immunoglobulin G (IgG). IL-21 influences the differentiation of NK cells by promoting the upregulation of Fc$\gamma$RIIIa (CD16). IL-21 also induces the secretion of IFN-$\gamma$ by NK cells.[40] IL-21 has been shown to augment NK cell perforin expression, proliferation, and degranulation.[41] NK cells exposed to IL-21 have a more mature phenotype and exhibit enhanced effector functions and reduced proliferation. IL-21, although a product of the adaptive immune response, induces a functionally mature stage of NK cells that prime the adaptive immune response and suppress the ongoing innate immune response. The effect of IL-21 on NK differentiation is dependent on whether the cell is resting or activated. IL-21 has been shown to block the recruitment of resting NK cells in response to the growth-promoting effects of IL-15. This block initiates a delayed apoptotic program for NK cells that are already activated and terminally differentiated.[28,42] IL-15 is essential for the ontogeny of NK cells whereas IL-21 plays an important role in early differentiation of NK cells and in acquisition of its effector functions.[43,44] Stimulation of NK cells with IL-21 downregulates the NKG2D activating receptor and therefore modifies target cell recognition by human NK cells.[7] IL-21 upregulates NKp30 and 2B4 receptors and also induces the expression of killer Ig-like receptors (KIRs), and CD2.[45]

IL-21 has been found to induce the activation of STAT proteins in human primary NK cells and the NK cell line NK-92. Although it has been reported that STAT3 is the primary STAT activated by IL-21, Roda *et al.*[40] used flow cytometric analysis to show that NK cells stimulated with IL-21 exhibited elevated intracellular levels of phosphorylated STAT1 (P-STAT1), P-STAT3, and P-STAT5 but not P-STAT2 or P-STAT4 in response to IL-21. In an *in vitro* study wild-type NK cells displayed synergistic IFN-γ production in response to immobilized IgG and IL-21, cells from STAT1-deficient mice were unable to secrete IFN-γ in response to IL-21 with or without immobilized IgG. Importantly, IFN-γ secretion in response to immobilized IgG alone, while low, was not affected in STAT1-deficient cells. Synergistic induction of IFN-γ following co-stimulation of NK cells with immobilized IgG and IL-21 was also lost following inhibition of Erk with U0126. Chemical inhibition of the MAPK family members' p38 and JNK did not diminish IFN-γ secretion by co-stimulated NK cells. These results suggest that cytokine secretion by NK cells in response to FcR and IL-21R activation is dependent upon distinct signaling cascades emanating from the IL-21R (P-STAT1 but not P-STAT4) and the FcγRIIIa (P-Erk). Co-administration of IL-21 (10 μg/mouse) and trastuzumab i.p. to immunocompetent mice bearing huHER2-positive tumors (CT-26[HER2/neu]) resulted in increased plasma levels of IFN-γ as compared to mice receiving IL-21 and control antibody

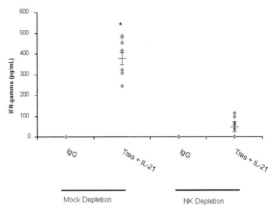

**Figure 1.** Co-administration of IL-21 and trastuzumab-coated, HER2-positive tumor cells to mice leads to NK cell IFN-γ production *in vivo*. Trastuzumab-coated CT-26HER2/neu murine tumor cells and IL-21 were coadministered to mice (*n* = 7) that had been depleted of NK cells via i.p. administration of anti-asialo GM1. Mock-depleted mice received injections of PBS. Control groups included mice that received tumor cells treated with normal huIgG (IgG). Serum was harvested from each mouse at 24 h post injection and analyzed for IFN-γ content by ELISA. *, *P* < 0.001 versus all conditions shown. (Reproduced with permission from The Journal of Immunology, 2006, 177: 120–129.)

treatment. IFN-γ secretion was markedly decreased when mice were depleted of NK cells prior to therapy (Fig. 1).

Cytokines can markedly and synergistically enhance the immune response to antibody-coated tumor cells *in vitro*, in murine models and in the context of translational phase I clinical trials.[40,46,47] Roda *et al.*[40,48] demonstrated that natural killer (NK) cells produce IFN-γ and multiple chemokines (e.g., MIP1α, TNFα, RANTES, IL-8) in response to tumor cell lines coated with an anti-HER1 monoclonal antibody (cetuximab) or an anti-HER2 monoclonal antibody (trastuzumab) and that this effector function was significantly enhanced in the presence of IL-21. This NK cell-derived IFN-γ markedly enhanced monocyte antibody-dependent cellular cytotoxicity (ADCC) and cytokine secretion, and NK cell-derived chemokines stimulated the migration of both naïve and activated T cells. NK cell ADCC against cetuximab-coated targets was

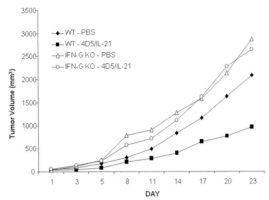

**Figure 2.** IL-21 enhances the effect of an antitumor monoclonal antibody in a murine tumor model. Wild-type or IFN-γ-deficient mice bearing s.c. CT-26HER2/neu tumors were treated i.p. with PBS or the combination of 4D5 (a murine monoclonal antibody recognizing human HER2) and IL-21. Tumor dimensions were measured, and tumor volumes were calculated as described.[40] SE was <5% for each data point shown. (Reproduced with permission from The Journal of Immunology, 2006, 177: 120–129.)

most effectively enhanced by co-administration of IL-21 as compared to IL-2 or IL-12. Importantly, the efficacy of cetuximab monoclonal antibody therapy in a HER1-positive murine lung cancer xenograft model was significantly enhanced when IL-21 was administered as an adjuvant. These results demonstrated that IL-21 can enhance the antitumor effects of therapeutic monoclonal antibodies by activating the innate immune system.

A murine tumor model was employed in order to determine whether IL-21 could enhance the effect of an anti-HER2/neu monoclonal antibody *in vivo*.[40] Wild-type or IFN-γ-deficient mice were inoculated s.c. with CT-26[HER2/neu] tumors and then treated with PBS or with murine anti-HER2 antibody (4D5) and IL-21 (Fig. 2). By day 23, wild-type mice receiving 4D5 and IL-21 exhibited an approximately 50% reduction in tumor volume as compared to wild-type mice receiving PBS, consistent with previous experiments ($P = 0.0005$). In contrast, tumor growth in the IFN-γ-deficient mice receiving 4D5 and IL-21 was not significantly different than tumor growth in the IFN-γ-deficient mice receiving PBS ($P = 0.6172$).

These data demonstrated that IL-21 can enhance the effect of a therapeutic monoclonal antibody in a murine solid tumor model, and that this effect is dependent on IFN-γ.

## Clinical Applications

Twenty-nine patients with metastatic melanoma received IL-21 in a phase I trial to study the safety of increasing doses. IL-21 was administered in various doses ranging from 1 to 100 μg/kg. Two parallel regimens were employed. Some patients received IL-21 three times per week (3/wk) for 6 weeks or three times a day for 5 days followed by 9 days of rest (5 + 9). IL-21 was generally well tolerated and no dose-limiting toxicities (DLTs) were observed at the 1, 3, or 10 μg/kg dose levels using these two regimens. The DLTs at higher IL-21 concentrations were similar in the two regimens tested. In the 3/wk regimen, toxicities included increases in liver enzymes, neutropenia, and lightheadedness with fever and rigors. DLTs in the 5 + 9 regimen included increases in liver enzymes, neutropenia, fatigue, and thrombocytopenia. The maximum tolerated dose of IL-21 was 30 μg/kg administered via i.v. bolus injection. On a 3/wk or 5 + 9 regimen IL-21 was found to be biologically active at all dose levels administered. These effects included increased levels of soluble CD25 indicative of NK cell and T cell activation and up-regulation of perforin and granzyme B mRNA in CD8[+] cells. One patient showed complete response, nine patients had stable disease, one showed a partial response, and 15 had progressive disease. Phase II studies commenced using 30 μg/kg and the 5 + 9 regimen.[49] In the phase II trial two patients with lung metastasis showed a clinical response – one partial and one complete.[50]

Encouraging results were observed in another phase I trial of IL-21 in patients with renal cell carcinoma or melanoma. Forty-four patients were evaluated in this phase I trial.

IL-21 was administered on days 1–5 and repeated after 9–16-day rest. Patient tumors were evaluated on day 28. IL-21 was tested at doses of 3, 10, 30, 50, and 100 μg/kg. 30 μg/kg of IL-21 was determined to be the maximum tolerable dose. Patients with tumor regression or stable disease were treated with the same dose for additional cycles. IL-21 was generally well tolerated with no grade 3 or 4 clinical toxicities with the first two cycles of therapy.[15,51,52] In metastatic melanoma patients, one patient showed a complete response and 11 had stable disease. In renal cell carcinoma four patients had partial responses and 13 had stable disease.

## Conclusions

IL-21 and its receptor IL-21R was first reported in the year 2000. In a short time, the effect of IL-21 has been very well studied, and it is already in clinical trials for several diseases. The biology of IL-21 is unique. IL-21 has pleotropic effects on a variety of immune cells depending on the cell type and their stage of development. IL-21 plays an important role in the differentiation and signaling events of both innate immune system and the adaptive immune system. IL-21 regulates the differentiation of almost all subsets of T cells. The effects of IL-21 on T cell response could be exploited in the vaccine setting. In the context of B cells, IL-21 promotes differentiation, induces apoptosis and has an immunosuppressive effect. IL-21 inhibits activation and maturation of DCs. In addition to regulating the differentiation of NK cells, activation of NK cells by IL-21 induces secretion of various proinflammatory cytokines and enhances their natural cytotoxicity against tumor cells. IL-21 functions as an effective antitumor agent. Success in preclinical and clinical trials has established IL-21 as a viable antitumor therapy. The efficacies of such antitumor agents could be enhanced by combination with therapeutic monoclonal antibodies.

## Conflicts of Interest

The authors declare no conflicts of interest.

## References

1. Parrish-Novak, J., S.R. Dillon, A. Nelson, *et al.* 2000. Interleukin 21 and its receptor are involved in NK cell expansion and regulation of lymphocyte function. *Nature* **408:** 57–63.
2. Parrish-Novak, J., D.C. Foster, R.D. Holly & C.H. Clegg. 2002. Interleukin-21 and the IL-21 receptor: novel effectors of NK and T cell responses. *J. Leukoc. Biol.* **72:** 856–863.
3. di Carlo, E., D. de Totero, T. Piazza, *et al.* 2007. Role of IL-21 in immune-regulation and tumor immunotherapy. *Cancer Immunol. Immunother.* **56:** 1323–1334.
4. Davis, I.D., K. Skak, M.J. Smyth, *et al.* 2007. Interleukin-21 signaling: functions in cancer and autoimmunity. *Clin. Cancer Res.* **13:** 6926–6932.
5. Spolski, R. & W.J. Leonard. 2008. Interleukin-21: basic biology and implications for cancer and autoimmunity. *Annu. Rev. Immunol.* **26:** 57–79.
6. Brady, J., Y. Hayakawa, M.J. Smyth & S.L. Nutt. 2004. IL-21 induces the functional maturation of murine NK cells. *J. Immunol.* **172:** 2048–2058.
7. Burgess, S.J., A.I. Marusina, I. Pathmanathan, *et al.* 2006. IL-21 down-regulates NKG2D/DAP10 expression on human NK and CD8+ T cells. *J. Immunol.* **176:** 1490–1497.
8. Strengell, M., S. Matikainen, J. Siren, *et al.* 2003. IL-21 in synergy with IL-15 or IL-18 enhances IFN-gamma production in human NK and T cells. *J. Immunol.* **170:** 5464–5469.
9. Wang, G., M. Tschoi, R. Spolski, *et al.* 2003. in vivo antitumor activity of interleukin 21 mediated by natural killer cells. *Cancer Res.* **63:** 9016–9022.
10. Ma, H.L., M.J. Whitters, R.F. Konz, *et al.* 2003. IL-21 activates both innate and adaptive immunity to generate potent antitumor responses that require perforin but are independent of IFN-gamma. *J. Immunol.* **171:** 608–615.
11. Di Carlo, E., A. Comes, A.M. Orengo, *et al.* 2004. IL-21 induces tumor rejection by specific CTL and IFN-gamma-dependent CXC chemokines in syngeneic mice. *J. Immunol.* **172:** 1540–1547.
12. Smyth, M.J., M.E. Wallace, S.L. Nutt, *et al.* 2005. Sequential activation of NKT cells and NK cells provides effective innate immunotherapy of cancer. *J. Exp. Med.* **201:** 1973–1985.
13. Takaki, R., Y. Hayakawa, A. Nelson, *et al.* 2005. IL-21 enhances tumor rejection through a NKG2D-dependent mechanism. *J. Immunol.* **175:** 2167–2173.

14. Comes, A., O. Rosso, A.M. Orengo, *et al*. 2006. CD25$^+$ regulatory T cell depletion augments immunotherapy of micrometastases by an IL-21-secreting cellular vaccine. *J. Immunol.* **176:** 1750–1758.

15. Thompson, J.A., B.D. Curti, B.G. Redman, *et al*. 2008. Phase I study of recombinant interleukin-21 in patients with metastatic melanoma and renal cell carcinoma. *J. Clin. Oncol.* **26:** 2034–2039.

16. Asao, H., C. Okuyama, S. Kumaki, *et al*. 2001. Cutting edge: the common gamma-chain is an indispensable subunit of the IL-21 receptor complex. *J. Immunol.* **167:** 1–5.

17. Gagnon, J., S. Ramanathan, C. Leblanc & S. Ilangumaran. 2007. Regulation of IL-21 signaling by suppressor of cytokine signaling-1 (SOCS1) in CD8(+) T lymphocytes. *Cell Signal.* **19:** 806–816.

18. Strengell, M., A. Lehtonen, S. Matikainen & I. Julkunen. 2006. IL-21 enhances SOCS gene expression and inhibits LPS-induced cytokine production in human monocyte-derived dendritic cells. *J. Leukoc. Biol.* **79:** 1279–1285.

19. Zeng, R., R. Spolski, E. Casas, *et al*. 2007. The molecular basis of IL-21-mediated proliferation. *Blood* **109:** 4135–4142.

20. Konforte, D., N. Simard & C.J. Paige. 2009. IL-21: An Executor of B Cell Fate. *J. Immunol.* **182:** 1781–1787.

21. Mehta, D.S., A.L. Wurster, M.J. Whitters, *et al*. 2003. IL-21 induces the apoptosis of resting and activated primary B cells. *J. Immunol.* **170:** 4111–4118.

22. Jin, H., R. Carrio, A. Yu & T.R. Malek. 2004. Distinct activation signals determine whether IL-21 induces B cell costimulation, growth arrest, or Bim-dependent apoptosis. *J. Immunol.* **173:** 657–665.

23. Mehta, D.S., A.L. Wurster & M.J. Grusby. 2004. Biology of IL-21 and the IL-21 receptor. *Immunol. Rev.* **202:** 84–95.

24. Wurster, A.L., V.L. Rodgers, A.R. Satoskar, *et al*. 2002. Interleukin 21 is a T helper (Th) cell 2 cytokine that specifically inhibits the differentiation of naive Th cells into interferon gamma-producing Th1 cells. *J. Exp. Med.* **196:** 969–977.

25. Korn, T., E. Bettelli, W. Gao, *et al*. 2007. IL-21 initiates an alternative pathway to induce proinflammatory T(H)17 cells. *Nature* **448:** 484–487.

26. Nurieva, R., X.O. Yang, G. Martinez, *et al*. 2007. Essential autocrine regulation by IL-21 in the generation of inflammatory T cells. *Nature* **448:** 480–483.

27. Coquet, J.M., K. Kyparissoudis, D.G. Pellicci, *et al*. 2007. IL-21 is produced by NKT cells and modulates NKT cell activation and cytokine production. *J. Immunol.* **178:** 2827–2834.

28. Kasaian, M.T., M.J. Whitters, L.L. Carter, *et al*. 2002. IL-21 limits NK cell responses and promotes antigen-specific T cell activation: a mediator of the transition from innate to adaptive immunity. *Immunity* **16:** 559–569.

29. Suto, A., A.L. Wurster, S.L. Reiner & M.J. Grusby. 2006. IL-21 inhibits IFN-gamma production in developing Th1 cells through the repression of Eomesodermin expression. *J. Immunol.* **177:** 3721–3727.

30. Eriksen, K.W., H. Søndergaard, A. Woetmann, *et al*. 2009. The combination of IL-21 and IFN-[alpha] boosts STAT3 activation, cytotoxicity and experimental tumor therapy. *Mol. Immunol.* **46:** 812–820.

31. Thedrez, A., C. Harly, A. Morice, *et al*. 2009. IL-21-mediated potentiation of antitumor cytolytic and proinflammatory responses of human V{gamma}9V{delta}2 T cells for adoptive immunotherapy. *J. Immunol.* **182:** 3423–3431.

32. Spolski, R. & W.J. Leonard. 2008. The Yin and Yang of interleukin-21 in allergy, autoimmunity and cancer. *Curr. Opin. Immunol.* **20:** 295–301.

33. Peluso, I., M.C. Fantini, D. Fina, *et al*. 2007. IL-21 counteracts the regulatory T cell-mediated suppression of human CD4$^+$ T lymphocytes. *J. Immunol.* **178:** 732–739.

34. Li, Y. & C. Yee. 2008. IL-21 mediated Foxp3 suppression leads to enhanced generation of antigen-specific CD8+ cytotoxic T lymphocytes. *Blood* **111:** 229–235.

35. Medzhitov, R. & C.A. Janeway Jr. 1997. Innate immunity: impact on the adaptive immune response. *Curr. Opin. Immunol.* **9:** 4–9.

36. Mende, I. & E.G. Engleman. 2005. Breaking tolerance to tumors with dendritic cell-based immunotherapy. *Ann. N. Y. Acad. Sci.* **1058:** 96–104.

37. Brandt, K., S. Bulfone-Paus, A. Jenckel, *et al*. 2003. Interleukin-21 inhibits dendritic cell-mediated T cell activation and induction of contact hypersensitivity in vivo. *J. Invest. Dermatol.* **121:** 1379–1382.

38. Brandt, K., S. Bulfone-Paus, D.C. Foster & R. Ruckert. 2003. Interleukin-21 inhibits dendritic cell activation and maturation. *Blood* **102:** 4090–4098.

39. Brandt, K., P.B. Singh, S. Bulfone-Paus & R. Ruckert. 2007. Interleukin-21: a new modulator of immunity, infection, and cancer. *Cytokine Growth Factor Rev.* **18:** 223–232.

40. Roda, J.M., R. Parihar, A. Lehman, *et al*. 2006. Interleukin-21 enhances NK cell activation in response to antibody-coated targets. *J. Immunol.* **177:** 120–129.

41. Strbo, N., L. de Armas, H. Liu, *et al*. 2008. IL-21 augments natural killer effector functions in chronically HIV-infected individuals. *Aids* **22:** 1551–1560.

42. Vosshenrich, C.A. & J.P. Di Santo. 2001. Cytokines: IL-21 joins the gamma(c)-dependent network? *Curr. Biol.* **11:** R175–177.

43. Vosshenrich, C.A., T. Ranson, S.I. Samson, *et al.* 2005. Roles for common cytokine receptor gamma-chain-dependent cytokines in the generation, differentiation, and maturation of NK cell precursors and peripheral NK cells in vivo. *J. Immunol.* **174:** 1213–1221.

44. Skak, K., K.S. Frederiksen & D. Lundsgaard. 2008. Interleukin-21 activates human natural killer cells and modulates their surface receptor expression. *Immunology* **123:** 575–583.

45. Sivori, S., C. Cantoni, S. Parolini, *et al.* 2003. IL-21 induces both rapid maturation of human CD34$^+$ cell precursors towards NK cells and acquisition of surface killer Ig-like receptors. *Eur J. Immunol.* **33:** 3439–3447.

46. Parihar, R., J. Dierksheide, Y. Hu & W.E. Carson. 2002. IL-12 enhances the natural killer cell cytokine response to Ab-coated tumor cells. *J. Clin. Invest.* **110:** 983–992.

47. Parihar, R., P. Nadella, A. Lewis, *et al.* 2004. A Phase I Study of Interleukin-12 with trastuzumab in patients with HER2 overexpressing malignancies: Correlation between response and sustained interferon-gamma production. *Clin. Cancer Res.* **10:** 5027–5037.

48. Roda, J.M., T. Joshi, J.P. Butchar, *et al.* 2007. The activation of natural killer cell effector functions by cetuximab-coated, epidermal growth factor receptor positive tumor cells is enhanced by cytokines. *Clin. Cancer Res.* **13:** 6419–6428.

49. Davis, I.D., B.K. Skrumsager, J. Cebon, *et al.* 2007. An open-label, two-arm, phase I trial of recombinant human interleukin-21 in patients with metastatic melanoma. *Clin. Cancer Res.* **13:** 3630–3636.

50. Davis, I.D., B. Brady, R.F. Kefford, *et al.* 2009. Clinical and biological efficacy of recombinant human interleukin-21 in patients with stage IV malignant melanoma without prior treatment: A phase IIa trial. *Clin. Cancer Res.* **15:** 2123–2129.

51. Curti, B.D. 2006. Immunomodulatory and antitumor effects of interleukin-21 in patients with renal cell carcinoma. *Expert Rev. Anticancer Ther.* **6:** 905–909.

52. Skak, K., M. Kragh, D. Hausman, *et al.* 2008. Interleukin 21: combination strategies for cancer therapy. *Nat. Rev. Drug Discov.* **7:** 231–240.

# Clinical and Immunologic Basis of Interferon Therapy in Melanoma

## Ahmad A. Tarhini and John M. Kirkwood

*University of Pittsburgh Cancer Institute, Pittsburgh, Pennsylvania, USA*

Interferon α2b (IFN-α2b) at high dosage is critical to the reversal of signaling defects in T cells of melanoma patients, and to the durable effector (α DC1) polarization of dendritic cells. These immunoregulatory effects appear to be uniquely achieved with levels of IFN-α only attainable *in vivo* using the high-dose regimen of IFN-α2b (HDI). Three US cooperative group studies have evaluated the benefit of HDI as an adjuvant therapy for high-risk melanoma. All have demonstrated significant and durable reduction in the frequency of relapse, while the first and third trials have demonstrated significant improvements in the fractions of patients surviving compared with observation (E1684) or with a ganglioside vaccine (GMK, E1694). A meta-analysis of 13 randomized trials evaluating adjuvant IFN therapy has now also demonstrated significant benefits for IFN in terms of RFS and OS. Research of IFN-α in melanoma is now focused on identifying prognostic markers of outcome and predictors of therapeutic response.

*Key words:* melanoma; interferon-α; adjuvant; neoadjuvant

## Introduction to Melanoma

Survival of melanoma varies widely by stage, from a highly curable disease when detected in early stages, to a disease with dismal prognosis when in advanced inoperable stages.[1] The American Joint Committee on Cancer (AJCC) divides cutaneous melanoma into four stages. Primary tumors confined to the skin without regional lymph node involvement are assigned stages I and II depending on the thickness (depth) of the tumor, and the presence or absence of ulceration of the overlying epithelium, or invasion of the reticular dermis or subcutaneous fat (Clark level IV or V). Stage III comprises a disease with clinical or pathological evidence of regional lymph node involvement, or the presence of in transit or satellite metastases. Stage IV disease is defined by the presence of distant metastasis. Patients with stage I melanoma have an excellent prognosis with surgical treatment alone and a cure rate of more than 85%. The 3–5 years postsurgical relapse rate in patients with stages IIA and IIB is 20–30% and 40–55% respectively. Stage III melanoma patients with regional lymph node involvement have a 5-year relapse rate of 40–80%, while stage IV disease has a dismal prognosis with a median survival of only 6 to 9 months.[2,3]

## Immunity in Melanoma and Implications for Adjuvant Immunotherapy

Immunity to melanoma appears to be important for disease control in the adjuvant and advanced disease settings. Spontaneous regression of disease has been reported in patients with melanoma, suggesting a role for host immunity, indirectly supported by the pathological evidence for the presence of lymphoid infiltrates at primary melanoma associated with tumor regression. Host cellular immune responses within melanoma have potential prognostic and predictive significance. T cell infiltrates in primary melanoma have been

Address for correspondence: Ahmad A. Tarhini, University of Pittsburgh, UPMC Cancer Pavilion, 5150 Centre Avenue, 5th floor, Pittsburgh, PA 15232. Voice: 412 648 6507; fax: 412 648 6579. tarhiniaa@upmc.edu suggested to be of prognostic significance,[4] and T cell infiltrates within regional nodal metastases predicts benefit in patients treated with neoadjuvant IFN-α2b therapy.[5–7]

The quality of the host immune response has been shown to differ between earlier and more advanced disease settings. While T helper type 1 (Th1)-type CD4$^+$ antitumor T-cell function appears to be critical to the induction and maintenance of antitumor cytotoxic T-lymphocyte (CTL) responses *in vivo*, and Th2- or Th3/Tr-type CD4$^+$ T-cell responses may subvert Th1-type cell mediated immunity yielding a microenvironment that facilitates disease progression, patients with active melanoma or renal cell carcinoma have been shown to display strong tumor antigen specific Th2-type polarization. On the other hand, normal donors and patients who were disease free following therapy demonstrate either strongly polarized mixed Th1-/Th2-type or str Th1-type responses to the same epitopes.[8,9] Therefore, host immune tolerance appears to be an impediment to the therapy of advanced disease, and this may be avoidable in the high-risk setting of operable disease, where the host susceptibility to immunologic interventions may be greater, and where IFN-α2b has demonstrated its significant impact upon melanoma relapse and survival.

## Interferon-α in the Treatment of Melanoma

IFN-α was the first recombinant cytokine to be investigated clinically for the therapy of metastatic melanoma. Initial phase I-II studies yielded overall response rates of about 16% and about one third achieved complete and durable responses. Responses were observed as late as six months from initiation of therapy, and up to one third of the responses were durable. However, the median duration of response was only about four months.[10–13] The use of IFN-α for adjuvant therapy of patients with melanoma is based on the hypothesis that

micrometastatic disease is the source of future relapse, and less established in its induction of host tolerance of tumor. While patients with advanced metastatic melanoma display immunological tolerance, the adjuvant setting may be more susceptible to interventions designed to induce Th1 host-effector mechanisms to eradicate micrometastases.

Multiple IFN-α2b regimens, that may be categorized as high-dose, intermediate-dose, or low-dose regimens, have been evaluated as adjuvant therapy for intermediate/high-risk (T3–4, lymph node positive) surgically resected melanoma (Table 1). The only randomized controlled trials that have shown durable relapse-free survival (RFS) and overall survival (OS) impact have utilized the high-dose IFN-α2b regimen (HDI).[14–16] A meta-analysis of 12 randomized trials of adjuvant IFN-α2b for high-risk melanoma confirmed a highly significant reduction in the odds of recurrence in patients treated with IFN compared to observation. The analysis also demonstrated evidence of increased benefit with increasing IFN dose and a trend for improved benefit with increasing total dose. This meta-analysis did not find a statistically significant OS benefit for IFN-α2b, although a larger individual patient data meta-analysis of 13 randomized trials showed a significant though small impact of IFN upon OS.[17,18] In this latter meta-analysis, there was statistically significant benefit for interferon (IFN) for both relapse-free survival (RFS) (OR = 0.87, CI = 0.81–0.93, $P = 0.00006$) and overall survival (OS) (0.9, 0.84–0.97, $P = 0.008$). This survival advantage translates into an absolute benefit of about 3% (CI 1%–5%) at 5 years. This analysis did not, however, clarify whether there is an optimal (high, intermediate or low) dose of IFN.[18]

## Adjuvant High-dose IFN-α2b (HDI) for High-risk Resected Melanoma

Since 1984, three national cooperative group studies have evaluated the benefit of high-dose

**TABLE 1.** Published Clinical Trials of Adjuvant IFN-α for Intermediate/High-risk (T3–4, Lymph Node Positive) Surgically Resected Melanoma

| Cooperative group/PI | Eligibility | n | Treatment agent/ dosage/duration | Impact on DFS | OS |
|---|---|---|---|---|---|
| NCCTG 837052 Creagan | T3–4, N1 | 262 | IFN-α2a 20 MU/m2/D IM TIW x3 mos | − | − |
| ECOG 1684 Kirkwood | T4, N1 | 287 | IFN-α2b 20 MU/m2/D IVx1 mo 10 MU/m2 SC TIW for 11 mos | + at 6.9–12.6 yrs | + |
| E1690 Intergroup Kirkwood | T4, N1 | 642 | IFN-α2b 20 MU/m2/D IVx1 mo 10 MU/m2 SC TIWx11 mos vs 3 MU/D SC TIWx2 yrs | + at 4.3– 6.6 yrs | − |
| WHO #16 Cascinelli | N1–2 | 444 | IFN-α2a 3 MU/D SC TIWx3 yrs | − | − |
| EORTC 18871 Kleeberg | T3–4, N1 | 830 | IFN-α2b 1 MU/D SC QODx1 yr vs IFNg 0.2 mg/D SC QODx1yr | − | − |
| E1694 Intergroup Kirkwood | T4, N1 | 880 | IFN-α2b 20 MU/m2/D IVx1 mo 10 MU/m2 SC TIWx11 mos vs GMK vaccine x 96 wks | + at 1.3–2.1 yrs | + |
| ECOG 2696 Kirkwood | T4, N1, M1 | 107 | GMK + IFN or –>IFN vs GMK | + at 1.4–2.6 yrs | − |
| UKCCR Aim-High Hancock | T4, N1 | 674 | IFN-α2a 3 MU/D SC QODx2 yrs | − | − |
| EORTC 18952 Eggermont | T4, N1–2 | 1488 | IFN-α2b 10MU/d then10 MU TIW x1 or 5 MU TIW x 2 years | +/− + | − |
| EORTC 18991 Eggermont | TxN1–2 | 1256 | Peg-IFN-α2b SC 6 μg/kg/week (8 weeks) then 3 μg/kg/week (5 years) vs Observation | + | − |
| French Grob | T2-T4 (≥1.5 mm), N0 | 489 | IFN-α2a 3 MU SC TIW x 18 mo vs Obs | ± | − |
| Austrian Pehamberger | T2-T4 (≥1.5 mm), N0 | 311 | IFN-α2a 3 MU SC QD x 3 wks then TIW x 1 yr vs Obs | ± | − |

IFN (HDI) as adjuvant therapy for resectable high-risk cutaneous melanoma. These included patients with regional lymph node metastases ($T_{1-4}$, $N_1$, $M_0$) and primary localized deep melanomas ($T_4$, $N_0$, $M_0$) that have a 5-year postsurgical relapse rate of more than 40–50%.

The first and third of these studies both demonstrated significant survival prolongation, compared to observation (E1684; The median RFS was 1.72 years in the high-dose interferon α-02b (HDI) arm versus 0.98 year in the Obs arm [stratified log-rank one-sided $P$ value ($P_1$) = 0.0023], and the median OS was

3.82 versus 2.78 years ($P_1 = 0.0237$), respectively)[14] and compared to a vaccine (GMK) that was selected as the optimal vaccine candidate at the time (E1694). The results of this trial were reported in 2001 based on a final analysis in June 2000, with a median follow-up interval of 16 months. Among eligible patients in this trial, HDI provided a statistically significant RFS benefit (HR = 1.47; $P_1 = 0.0015$) and OS benefit (HR = 1.52; $P_1 = 0.009$) compared with GMK. A similar benefit was observed in the intent-to-treat analysis of RFS (HR = 1.49) and OS (HR = 1.38).[19]

The second trial, E1690, conducted in part before and in part after the approval of HDI, was associated with systematic crossover of patients from the observation-assigned arm to treatment at nodal relapse with HDI. This trial showed differences in terms of RFS but not OS. In the intent-to-treat analysis of RFS, treatment with HDI was associated with a statistically significant benefit compared with Obs (HR = 1.28; $P_1 = 0.025$). In contrast, LDI was not associated with a significant RFS benefit compared with Obs. Neither HDI nor LDI regimens had any apparent impact on OS compared with Obs in this trial. However, a retrospective analysis of salvage therapy demonstrated the occurrence of a disproportionate crossover of patients from the Obs arm to HDI therapy postprotocol in those patients who developed regional recurrence (stage IIB patients in this trial were not required to undergo lymphadenectomy), which may have confounded the survival analysis.[15]

The analysis of each of the foregoing studies has been conducted at the closure of each study[14,15,19] and has been updated in a pooled analysis of survival and relapse-free outcomes to April 2001.[20] The pooled analysis has firmly demonstrated that melanoma relapse has been prevented by IFN to intervals that now approach 20 years, and yet this analysis has not yielded compelling evidence of an impact upon OS despite the positive survival results of two randomized US Cooperative Group and Intergroup studies

(E1684 and E1694). This may not be surprising, given that the larger of the two observation-controlled trials included in the pooled analysis (E1690) did not show an OS benefit for HDI. As discussed previously, the confounding of the OS analysis of E1690 by the routine crossover to HDI of all but one of 37 patients assigned to observation who had nodal relapse, associated with an unusually prolonged postrelapse survival of those patients in the observation arm treated with HDI, may have been responsible for this outcome variability. Patients treated with HDI in E1694 have not been included in the pooled analysis because the comparator in that trial was the GMK vaccine and not observation as was the case in E1684 and E1690.

An interim analysis of a phase III trial (EORTC 18961) of adjuvant GMK vaccination versus observation after resection of the primary in AJCC stage II (T3-T4, N0, M0) melanoma patients was recently reported.[21] For the primary endpoint, RFS, the criteria for stopping for futility were met. For OS, the results suggested a detrimental effect of the vaccine, although this trial is so early in its follow-up that it is difficult to interpret this finding given that no adverse effect was seen in terms of the generally more sensitive endpoints of RFS, and distant metastasis-free survival. The investigators concluded that GMK vaccination is ineffective and might even be detrimental in the stage II melanoma adjuvant setting. These data have led some investigators to question the evidence for a benefit of HDI upon OS in high-risk melanoma patients studied in E1694. On the other hand, there continues to be a wide agreement upon the RFS benefit of HDI supported by all three randomized trials (E1684, E1690, E1694) as well as the data from multiple lower-dose IFN trials summarized by Wheatley and Ives.[22] As discussed earlier, this individual patient data meta-analysis of 13 randomized trials showed a significant though small impact of IFN upon OS (HR = 0.9 with an absolute 5-year increment of 3%). In addition, no other agent has ever been demonstrated to provide similar relapse-free or survival benefits for this patient population. Therefore, HDI continues to be the current standard and the only option available for these patients outside of a clinical trial.

## The Role of Dose, Route, and Duration of IFN-α Therapy in Melanoma

Less intensive (less toxic) regimens tested in the melanoma adjuvant setting have not previously ever demonstrated durable effects upon relapse or death as has been observed with HDI. These include very low-dose IFN (1 MU SC QOD) as tested in EORTC 18871 (AJCC TNM melanoma stage T3–4, N1)[23] and low-dose IFN (3 MU SC TIW) as tested in WHO Trial 16 (N1–2),[24] ECOG 1690 (T4, N1),[15] UKCCR AIM-High trial (T4, N1),[25] and the Scottish trial.[26] These also include intermediate-dose IFN (given subcutaneously) regimens tested in EORTC 18952 (T4, N1–2)[27] and EORTC 18961 (TxN1–2).[21]

All trials of IFN-α with durable RFS and OS impact utilized an intravenous (i.v.) induction phase given at 20MU/m2 five days a week for four weeks ($C_{max} > 10,000$ u/ml). Based on this experience, a US intergroup trial, E1697 was designed to evaluate the impact of a 4-week course of high dose IFN-α2b similar to the induction phase of the HDI regimen tested in E1684/E1690/E1694 trials. Eligibility for this trial includes patients with resected melanoma in the following categories: (1) $T_{2b} N_0$ (2) $T_{3a-b} N_0$ (3) $T_{4a-b} N_0$ (4) $T_{1-4} N_{1a,2a}$ (microscopic), and the control arm is observation. Out of the 1420 patients planned for accrual (to demonstrate a 7% improvement in RFS) to this study, more than 950 have been accrued as of April 2009. Other groups are also testing the role of the I.V. induction phase of HDI, including the Italian Melanoma Intergroup (HDI Induction given every 2 months for a total of four courses; 80 doses, $n = 300$) and DeCOG (HDI Induction given every 4 months for one year; 60 doses, $n = 800$).

The Hellenic Oncology Group conducted a randomized, phase 3 trial to evaluate intravenous induction therapy with IFN-α2b for 4 weeks as compared with the same regimen followed by 11 months of adjuvant IFN-α2b therapy. This trial used a modified high-dose adjuvant IFN regimen derived from the E1684/1690/1694 HDI regimen. The study design proposed that the one-month treatment would be considered at least as good as the one-year regimen treatment, if the relapse rate at 3 years from study entry is at most 15% higher in the former arm. A relapse rate of 60% was assumed in both treatment arms. A sample size of 152 (182 enrolled) patients per treatment arm was planned. The trial concluded that at the 5% level of significance the 3-year relapse rate of the one month group was not 15% higher than the relapse rate of the one-year group. There were also no significant differences in OS, disease-free survival (DFS) or severe toxicities between the arms. The issues with this study are related to the adoption of a nonstandard IFN induction dose regimen for one month (15 Mu/M2 compared to the 20 Mu/M2 HDI dosing; $n = 182$) and a nonstandard $3/4$ IFN maintenance regimen for the balance of one year (delivering 10 Mu/dose rather than 10 Mu/M2 as specified in the FDA-approved HDI dosage regimen; $n = 182$). This study arm of one month would have termed as noninferior, a treatment with 15% lower 3-year RFS/OS. Beyond this, in the absence of an observation control it is not certain what level of activity was achieved by either arm. By comparison, a US cooperative group proposal to test equivalence of one month and one year (with a 5% threshold) was designed and abandoned when it became apparent that it would have required 3000 patients in 1991.

A second major question that has not been clearly answered is whether prolonged adjuvant therapy may offer improved results. A Dermatologic Cooperative Oncology Group (DeCOG) study of lower-dose IFN-α2a treatment at 3 MU 3x/wk s.c. for 5 years versus 18 months showed no difference in RFS or OS.[28]

The EORTC 18952 trial tested two intermediate dosages of IFN (IDI) administered over 2 years versus one year and showed nonsignificant differences in favor of the 2-year regimen.[29] The Nordic IFN-α trial comparing an identical regimen for 2 years versus one year of IDI showed nonsignificant differences in favor of the one-year regimen. Most recently, the EORTC trial 18991 testing Peg-IFN-α has shown neither improved OS nor DMFS overall, although RFS benefits overall have been noted on analysis for regulatory review, and these appear to be confined to the subset of patients without gross nodal disease (termed N1 by the EORTC). The RFS difference of 16% at 5 years was significant in the N1 group with microscopic nodal disease. Unfortunately, this trial designed to deliver 5 years of therapy has shown a median treatment interval of only one year, so the question of whether longer therapy with this regimen achieves more significant antitumor effects can not be answered at this time. The hypothesis that was being tested, that prolonged lower-dose therapy might achieve anti-angiogenic effects has not yet been analyzed.

The meta-analysis of individual patient data that was undertaken by Ives, Wheatley and the leaders of the cooperative group trials of IFN through 2007 initially suggested that the presence of primary tumor ulceration predicted increased susceptibility to IFN therapeutic effects.[30] A more recent analysis of the adjuvant trials EORTC18952 and EORTC18991 has assessed the predictive value of ulceration in relation to the therapeutic impact of IFN-α in terms of RFS, DMFS and OS, overall, and according to stage: IIB and III (N1 microscopic nodal, N2 macroscopic nodal disease). Among 2,644 patients randomized into these studies less than one-third (849) had ulcerated primaries, and 1,336 nonulcerated primaries, while for 459 the ulceration status was unknown. In the group of patients with ulcerated primary melanomas, the impact of IFN was noted to be greater than in the nonulcerated group for RFS (test for interaction: $P = 0.02$), DMFS ($P < 0.001$), and OS ($P < 0.001$). The greatest effects of therapy were noted in patients with ulceration and stages IIB/III-N1. Based on this retrospective analysis, the EORTC have planned the EORTC 18081 trial, which will compare the benefit of Peg-IFN-α2b versus observation in patients with ulcerated primaries and Breslow depth of more than 1mm (node negative). It is noteworthy that unlike US cooperative groups, the EORTC does not require central pathology review for EORTC melanoma trials.[31]

## Neoadjuvant HDI Treatment of Potentially Resectable Local-Regional Metastases of Cutaneous Melanoma

Patients with clinically palpable regional lymph node metastases (AJCC stage IIIB-C; $T_{any}$, $N_{1b,2b,2c,3}$) carry a risk of relapse and death that approaches 70% at 5 years.[1,32,33]

In a pilot study, neoadjuvant HDI was investigated in this group of patients, who underwent surgical biopsy at study entry and then received standard intravenous HDI (20 million units/m$^2$, 5 days per week) for 4 weeks followed by complete lymphadenectomy. This was followed by standard maintenance subcutaneous HDI (10 million units/m$^2$ three times per week) for 48 weeks. Biopsy samples were obtained before and after intravenous HDI and subjected to immunohistochemical (IHC) analysis as well as routine pathologic study. Twenty patients were enrolled, and biopsy samples were informative for 17. Eleven patients (55%) demonstrated objective clinical response, and three patients (15%) had complete pathologic response. At a median follow-up of 18.5 months (range, 7–50 months) 10 patients had no evidence of recurrent disease.[34] In the context of this neoadjuvant study, HDI was found to upregulate pSTAT1, whereas it downregulates pSTAT3 and total *STAT3* levels in both tumor cells and lymphocytes. Higher pSTAT1/pSTAT3 ratios in tumor cells pretreatment were associated with longer OS ($P = 0.032$). The pSTAT1/pSTAT3 ratios were augmented by HDI both in melanoma cells ($P = 0.005$) and in lymphocytes ($P = 0.022$). Of the immunologic mediators and markers tested, *TAP2* was augmented by HDI (but not *TAP1* and MHC class I/II).[35] In addition, HDI was found to regulate MAPK signaling differentially in melanoma tumor cells and host lymphoid cells. HDI was found to downregulate pSTAT3 ($P = 0.008$) and phospho-MEK1/2 ($P = 0.008$) levels significantly in tumor cells. Phospho-ERK1/2 was downregulated by HDI in tumor cells ($P = 0.015$), but not in lymphoid cells. HDI downregulated *EGFR* ($P = 0.013$), but pSTAT3 activation appeared not to be associated with *EGFR* expression and the MEK/ERK MAPK pathway, indicating that *STAT3* activation is independent of the EGFR/MEK/ERK signaling pathway[36] Clinical responders had significantly greater increases in endotumoral CD11c$^+$ and CD3$^+$ cells and significantly greater decreases in endotumoral CD83$^+$ cells compared with nonresponders.[7]

# Improved Survival and Relapse-free Interval Associated with Autoimmunity Induced by HDI Therapy for High-risk Melanoma

Recent studies of immunotherapy for melanoma including high-dose IL-2,[37] anti-CTLA4 antibody[38-40] have suggested a correlation of antitumor effects and autoimmune phenomena like thyroiditis, hypophysitis, enteritis, hepatitis, and dermatitis. More recently, patients who have shown a strong correlation of prolonged RFS and OS after treatment with the modified adjuvant IFN regimen (Hellenic) related to the E1684/1690/1694 HDI regimen have demonstrated a strong correlation with autoimmune phenomena and/or the appearance of auto-antibodies in the serum.[41] Autoantibodies were detected in 52 (26%) of the group of 200 patients tested. Clinical manifestations of autoimmunity were observed among 15 (7%) of patients including vitiligo-like depigmentation in 11 (5%). A total of 113 patients have relapsed and 82 have died. The median time to progression (TTP) was 27.6 months and the median survival was 58.6 months. The median TTP for the patients who did not develop clinical or serological evidence of autoimmunity was 15.9 months while it has not been reached for the 52 patients who developed autoimmunity (106 vs 7; $P < 0.0001$). The median survival was 37.5 months for those who were negative and has not been reached for the other group (80 vs. 2 $P < 0.001$). In multivariate analysis the presence of autoimmunity was an independent favorable prognostic marker.

We evaluated the E2696 and E1694 trials to better understand the prognostic value of autoimmunity induced by HDI. In E2696, patients with resectable high risk melanoma were randomized to GM2-KLH/QS-1 (GMK) vaccine plus concurrent HDI, GMK plus sequential HDI, or GMK alone. In E1694, patients were randomized to either GMK or HDI. Sera from 103 patients in E2696 and 691 patients in E1694 banked at baseline and up to three additional time points were tested by ELISA for the development of five autoantibodies. In E2696, autoantibodies were induced in 17 subjects (25%; $n = 69$) receiving HDI and GMK versus 2 (6%; $n = 34$) receiving GMK without HDI (2$P$-value $= 0.029$). Of 691 patients in E1694, 67 subjects (19.3%; $n = 347$) who received IFN developed autoantibodies versus only 15 (4.4%; $n = 344$) in the vaccine control group (2$P$-value $< 0.001$). In the HDI arms, almost all induced autoantibodies were detected at $\geq 12$ weeks after initiation of therapy. A one-year landmark analysis of E1694 resected stage III patients, showed survival advantage associated with HDI-induced autoimmunity that approaches statistical significance (HR $= 1.54$; $P = 0.072$) adjusting for treatment.[42,43]

These observations support the hypothesis that prevention of melanoma relapse and mortality with IFN is associated with immunomodulation that may increase resistance to melanoma. IFN-α2b induction of autoimmunity may provide a useful surrogate biomarker of adjuvant therapeutic benefit. Studies of autoimmunity and its genetic determinants may help identify patients most likely to benefit from HDI and other newer immunotherapies associated with autoimmunity, such as the anti-CTLA-4 blocking antibodies ipilimumab and tremelimumab.

## Baseline Proinflammatory Cytokine Levels Predict Relapse-free Survival Benefit with HDI

The detection of serum biomarkers that are either prognostic of clinical outcome, or predictive of response to IFN-α2b has been pursued using high-throughput xMAP® multiplex immunobead assay technology (Luminex Corp.). This technology was utilized to simultaneously measure the levels of 29 cytokines, chemokines, angiogenic and growth factors, as well as soluble receptors in the sera of 179 patients with high-risk melanoma who have participated in the E1694, and 378 healthy age and gender-matched controls. These banked sera that have been tested were prospectively collected in the course of the intergroup E1694 trial. The 179 melanoma patients were chosen at random from the two trial arms according to disease status (whether the subject had relapsed at <1 year, between one and 3 years, or more than 5 years). Of those samples tested, 93 were derived from patients who received GMK vaccination and 86 were derived from patients treated with HDI. The clinical data from the E1694 trial were then mature to a median of 4.6 years of follow-up.

The results demonstrated that serum concentrations of IL-1α, IL-1β, IL-6, IL-8, IL-12p40, IL-13, G-CSF, MCP-1, MIP-1α, MIP-1β, IFN-α, TNF-α, EGF, VEGF, and TNFRII are significantly higher among patients with resected high-risk melanoma, when compared to healthy controls. Serum levels of immunosuppressive, angiogenic/growth stimulatory factors (VEGF, EGF, HGF) were decreased by IFN-α2b therapy significantly while levels of anti-angiogenic IP-10 and IFN-α were elevated post treatment. Comparing patients according to relapse outcome, the pretreatment levels of pro-inflammatory cytokines IL-1β, IL-1α, IL-6, TNF-α, and chemokines MIP-1α, and MIP-1β were significantly higher in sera of patients with longer RFS of greater than 5 years, compared with patients who experienced shorter RFS of less than one year.

## Conclusion

IFN-α2b at high dosage has demonstrated durable adjuvant therapeutic impact upon melanoma patients that are at high risk for recurrence and death after surgical resection, with either deep primary melanoma alone, or nodal involvement by microscopic or macroscopic tumor. An immunological mechanism of action has been suggested for the therapeutic actions of IFN-α2b and the reversal of signaling defects in T cells of melanoma patients and to the polarization of dendritic cells. The induction of autoimmunity is a marker of altering immunologic tolerance and may be utilized as a post-treatment correlate of therapeutic benefit. A baseline proinflammatory serum cytokine profile (IL-1α, β; MIP-1α, β; IL-6, TNF-α) pretherapy is predictive of benefit from HDI at 5 years. The restoration of T cell STAT1 signaling has also been advanced as a predictive tool for the selection of patients for therapy. In addition, biological features of the primary tumor such as ulceration have been advanced as potential tools for the selection of patients with a greater likelihood of benefit from lower-dose regimens of therapy, although the role of ulceration of the primary has not proven to be of use in relation to the FDA-approved regimen of HDI. These will be tested prospectively over the next 5 years. Ultimately, genetic factors that may distinguish patients who are susceptible to the immunoregulatory effects of IFNs and other immunotherapies are being sought in relation to current US intergroup trial E1697 that may be informative in the next few years.

## Conflicts of Interest

John Kirkwood: Schering Plough Speaker's Board.

## References

1. Balch, C.M., A.C. Buzaid, S.J. Soong, *et al.* 2001. Final version of the American Joint Committee on Cancer staging system for cutaneous melanoma. *J. Clin. Oncol.* **19:** 3635–3648.

2. Manola, J., M. Atkins, J. Ibrahim, *et al.* 2000. Prognostic factors in metastatic melanoma: a pooled analysis of Eastern Cooperative Oncology Group trials. *J. Clin. Oncol.* **18:** 3782–3793.

3. Kirkwood, J.M. & S.S. Agarwala. 1993. Systemic cytotoxic and biologic therapy of melanoma. In *Cancer: Principles and Practice of Oncology.* V.T. DeVita, S. Hellman & S.A. Rosenberg, Eds.: 1–16. Lippincott. Philadelphia, PA.

4. Clemente, C.G., M.C. Mihm Jr., R. Bufalino, *et al.* 1996. Prognostic value of tumor infiltrating lymphocytes in the vertical growth phase of primary cutaneous melanoma. *Cancer* **77:** 1303–1310.

5. Hakansson, A., B. Gustafsson, L. Krysander, *et al.* 1996. Tumour-infiltrating lymphocytes in metastatic malignant melanoma and response to interferon alpha treatment. *Br. J. Cancer* **74:** 670–676.

6. Mihm, M.C., C.G. Clemente & N. Cascinelli. 1996. Tumor infiltrating lymphocytes in lymph node melanoma metastases: a histopathologic prognostic indicator and an expression of local immune response. *Lab. Invest.* **74:** 43–47.

7. Moschos, S.J., H.D. Edington, S.R. Land, *et al.* 2006. Neoadjuvant treatment of regional stage IIIB melanoma with high-dose interferon alfa-2b induces objective tumor regression in association with modulation of tumor infiltrating host cellular immune responses. *J. Clin. Oncol.* **24:** 3164–3171.

8. Tatsumi, T., L.S. Kierstead, E. Ranieri, *et al.* 2002. Disease-associated bias in T helper type 1 (Th1)/Th2 CD4(+) T cell responses against MAGE-6 in HLA-DRB10401(+) patients with renal cell carcinoma or melanoma. *J. Exp. Med.* **196:** 619–628.

9. Tatsumi, T., C.J. Herrem, W.C. Olson, *et al.* 2003. Disease stage variation in CD4$^+$ and CD8$^+$ T-cell reactivity to the receptor tyrosine kinase EphA2 in patients with renal cell carcinoma. *Cancer Res.* **63:** 4481–4489.

10. Creagan, E.T., D.L. Ahmann, S. Frytak, *et al.* 1986. Phase II trials of recombinant leukocyte A interferon in disseminated malignant melanoma: results in 96 patients. *Cancer Treat. Rep.* **70:** 619–624.

11. Creagan, E.T., D.L. Ahmann, S. Frytak, *et al.* 1986. Recombinant leukocyte A interferon (rIFN-alpha A) in the treatment of disseminated malignant melanoma. Analysis of complete and long-term responding patients. *Cancer* **58:** 2576–2578.

12. Creagan, E.T., J.S. Kovach, H.J. Long, *et al.* 1986. Phase I study of recombinant leukocyte A human interferon combined with BCNU in selected patients with advanced cancer. *J. Clin. Oncol.* **4:** 408–413.

13. Creagan, E.T., A.J. Schutt, H.J. Long, *et al.* 1986. Phase II study: the combination DTIC, BCNU, actinomycin D, and vincristine in disseminated malignant melanoma. *Med. Pediatr. Oncol.* **14:** 86–87.

14. Kirkwood, J.M., M.H. Strawderman, M.S. Ernstoff, *et al.* 1996. Interferon alfa-2b adjuvant therapy of high-risk resected cutaneous melanoma: the Eastern Cooperative Oncology Group Trial EST 1684. *J. Clin. Oncol.* **14:** 7–17.

15. Kirkwood, J.M., J.G. Ibrahim, V.K. Sondak, *et al.* 2000. High- and low-dose interferon alfa-2b in high-risk melanoma: first analysis of intergroup trial E1690/S9111/C9190. *J. Clin. Oncol.* **18:** 2444–2458.

16. Kirkwood, J.M., J. Ibrahim, D.H. Lawson, *et al.* 2001. High-dose interferon alfa-2b does not diminish antibody response to GM2 vaccination in patients with resected melanoma: results of the Multicenter Eastern Cooperative Oncology Group Phase II Trial E2696. *J. Clin. Oncol.* **19:** 1430–1436.

17. Wheatley, K., N. Ives, B. Hancock, *et al.* 2003. Does adjuvant interferon-alpha for high-risk melanoma provide a worthwhile benefit? A meta-analysis of the randomised trials. *Cancer Treat. Rev.* **29:** 241–252.

18. Wheatley, K., N. Ives, A. Eggermont & J.M. Kirkwood. 2007. Interferon-alfa as adjuvant therapy for melanoma: an individual patient data meta-analysis of randomised trials, ASCO Annual Meeting. Chicago.

19. Kirkwood, J.M., J.G. Ibrahim, J.A. Sosman, *et al.* 2001. High-dose interferon alfa-2b significantly prolongs relapse-free and overall survival compared with the GM2-KLH/QS-21 vaccine in patients with resected stage IIB-III melanoma: results of intergroup trial E1694/S9512/C509801. *J. Clin. Oncol.* **19:** 2370–2380.

20. Kirkwood, J.M., J. Manola, J. Ibrahim, *et al*. 2004. A pooled analysis of eastern cooperative oncology group and intergroup trials of adjuvant high-dose interferon for melanoma. *Clin. Cancer Res.* **10:** 1670–1677.

21. Eggermont, A., S. Suciu & W. Ruka. 2008. EORTC 18961: Post-operative adjuvant ganglioside GM2-KLH21 vaccination treatment vs observation in stage II (T3-T4N0M0) melanoma: 2nd interim analysis led to an early disclosure of the results. *J. Clin. Oncol.* **26:** (May 20 suppl; abstr 9004), 2008 ASCO Annual Meeting Chicago.

22. Wheatley, K., N. Ives, A. Eggermont, *et al.*; International Malignant Melanoma Collaborative Group. 2007. Interferon-α as adjuvant therapy for melanoma: An individual patient data meta-analysis of randomised trials. In Oncology JoC, Ed.: Journal of Clinical Oncology: ASCO Annual Meeting Chicago.

23. Kleeberg, U.R., S. Suciu, E.B. Brocker, *et al*. 2004. Final results of the EORTC 18871/DKG 80-1 randomised phase III trial. rIFN-alpha2b versus rIFN-gamma versus ISCADOR M versus observation after surgery in melanoma patients with either high-risk primary (thickness >3 mm) or regional lymph node metastasis. *Eur. J. Cancer* **40:** 390–402.

24. Cascinelli, N., F. Belli, R.M. MacKie, *et al*. 2001. Effect of long-term adjuvant therapy with interferon alpha-2a in patients with regional node metastases from cutaneous melanoma: a randomised trial. *Lancet* **358:** 866–869.

25. Hancock, B.W., K. Wheatley, S. Harris, *et al*. 2004. Adjuvant interferon in high-risk melanoma: the AIM HIGH Study–United Kingdom Coordinating Committee on Cancer Research randomized study of adjuvant low-dose extended-duration interferon Alfa-2a in high-risk resected malignant melanoma. *J. Clin. Oncol.* **22:** 53–61.

26. Cameron, D.A., M.C. Cornbleet, R.M. Mackie, *et al*. 2001. Adjuvant interferon alpha 2b in high risk melanoma – the Scottish study. *Br. J. Cancer* **84:** 1146–1149.

27. Eggermont, A.M., S. Suciu, R. MacKie, *et al*. 2005. Post-surgery adjuvant therapy with intermediate doses of interferon alfa 2b versus observation in patients with stage IIb/III melanoma (EORTC 18952): randomised controlled trial. *Lancet* **366:** 1189–1196.

28. Hauschild, A., M. Volkenandt, W. Tilgen, *et al*. 2008. *Efficacy of interferon alpha 2a in 18 versus 60 months of treatment in patients with primary melanoma of 1.5 mm tumor thickness: A randomized phase III DeCOG trial*, pp. 15S. American Society of Clinical Oncology. Chicago, IL.

29. Eggermont, A.M., S. Suciu, M. Santinami, *et al*. 2008. Adjuvant therapy with pegylated interferon alfa-2b versus observation alone in resected stage III melanoma: final results of EORTC 18991, a randomised phase III trial. *Lancet* **372:** 117–126.

30. Ives, N.J., R.L. Stowe, P. Lorigan & K. Wheatley. 2007. Biochemotherapy versus chemotherapy for metastatic malignant melanoma: A meta-analysis of the randomised trials. ASCO Annual Meeting.

31. Eggermont, M., S. Suciu, A. Testori, *et al*. 2009. Ulceration of primary melanoma and responsiveness to adjuvant interferon therapy: Analysis of the adjuvant trials EORTC18952 and EORTC18991 in 2,644 patients. ASCO Annual Meeting. Orlando, FL.

32. Balch, C.M., S.J. Soong, T.M. Murad, *et al*. 1981. A multifactorial analysis of melanoma: III. Prognostic factors in melanoma patients with lymph node metastases (stage II). *Ann. Surg.* **193:** 377–388.

33. Balch, C.M., S.J. Soong, J.E. Gershenwald, *et al*. 2001. Prognostic factors analysis of 17,600 melanoma patients: validation of the American Joint Committee on Cancer melanoma staging system. *J. Clin. Oncol.* **19:** 3622–3634.

34. Kirkwood, J.M., M.S. Ernstoff, C.A. Davis, *et al*. 1985. Comparison of intramuscular and intravenous recombinant alpha-2 interferon in melanoma and other cancers. *Ann. Intern. Med.* **103:** 32–36.

35. Wang, W., H.D. Edington, U.N. Rao, *et al*. 2007. Modulation of signal transducers and activators of transcription 1 and 3 signaling in melanoma by high-dose IFNalpha2b. *Clin. Cancer Res.* **13:** 1523–1531.

36. Wang, W., H.D. Edington, U.N. Rao, *et al*. 2008. Effects of high-dose IFNalpha2b on regional lymph node metastases of human melanoma: modulation of STAT5, FOXP3, and IL-17. *Clin. Cancer Res.* **14:** 8314–8320.

37. Liu, K. & S.A. Rosenberg. 2003. Interleukin-2-independent proliferation of human melanoma-reactive T lymphocytes transduced with an exogenous IL-2 gene is stimulation dependent. *J. Immunother.* **26:** 190–201.

38. Ribas, A. 2005. Phase I trial of monthly doses of the human anti-CTLA4 monoclonal antibody CP-675, 206 in patients with advanced melanoma. Proc. ASCO, pp. 716.

39. Phan, G.Q. 2003. Cancer regression and autoimmunity induced by cytotoxic T lymphocyte-associated antigen 4 blockade in patients with metastatic melanoma. *Proc. Natl. Acad. Sci. USA.* **100:** 8372–8377.

40. Ribas, A., L.H. Camacho, G. Lopez-Berestein, *et al*. 2005. Antitumor activity in melanoma and anti-self responses in a phase I trial with the anti-cytotoxic T lymphocyte-associated antigen 4 monoclonal antibody CP-675,206. *J. Clin. Oncol.* **23:** 8968–8977.

41. Gogas, H., J. Ioannovich, U. Dafni, *et al.* 2006. Prognostic significance of autoimmunity during treatment of melanoma with interferon. *N. Engl. J. Med.* **354:** 709–718.

42. Tarhini, A.A., J. Stuckert, S. Lee, *et al.* 2007. Prognostic significance of serial serum S100 protein levels in high-risk surgically resected melanoma in ECOG phase II trial E2696, AACR. Los Angeles.

43. Stuckert, J., A.A. Tarhini, S. Lee, *et al.* 2007. Interferon alfa-induced autoimmunity in patients with high-risk melanoma participating in ECOG trial E2696, AACR. Los Angeles.

# Heterogeneous, Longitudinally Stable Molecular Signatures in Response to Interferon-β

## M. R. Sandhya Rani,[a] Yaomin Xu,[c] Jar-chi Lee,[c] Jennifer Shrock,[a] Anupama Josyula,[a] Joerg Schlaak,[d] Swathi Chakraborthy,[c] Nie Ja,[c] Richard M. Ransohoff,[a,b] and Richard A. Rudick[b]

[a]Neuroinflammation Research Center, Department of Neurosciences, [b]Mellen Center for MS Treatment and Research, Neuroscience Institute, [c]Quantitative Health Science, Lerner Research Institute, Cleveland Clinic, Cleveland, Ohio, USA

[d]University of Essen, Essen, Germany

Interferons (IFNs) are widely used in therapy for viral, neoplastic, and inflammatory disorders, but clinical response varies among patients. The biological basis for variable clinical response is not known. We determined the primary molecular response to IFN-beta (IFN-β) injections in 35 treatment-naïve multiple sclerosis (MS) patients using a customized cDNA macroarray with 186 interferon-stimulated genes (ISGs). Our results revealed striking interindividual heterogeneity, both in the magnitude as well as the nature of the primary molecular response to IFN-β injections. Despite marked between-subject variability in the molecular response, responses within individual subjects were stable over a 6-month interval. Our data suggest that clinical response to IFN-β therapy for MS differs among patients because of qualitative rather than quantitative variability in the primary molecular response to the drug.

*Key words:* Interferon-β; inflammatory; multiple sclerosis

## Introduction

Gene expression analysis is considered highly promising for the identification of biomarkers for predictive management of disease. It is hoped that specific patterns of gene expression can be used to characterize different types of disease or response to therapy.

Multiple sclerosis (MS) is an inflammatory neurodegenerative disease of unknown etiology. Biochemical tests for response to treatment are entirely lacking. Recombinant IFN-β was found efficacious based on empirical clinical trials.[1] The trials were not based on detailed understanding of MS pathogenesis, or of the likely mechanisms of action exerted by IFN-β. Nevertheless, IFN-β was shown to reduce relapses, new MRI lesions, and disability progression and is now standard therapy for patients with relapsing remitting MS. Despite worldwide use of IFN-β for MS, molecular mechanisms related to clinical benefits and toxicity are not known.

Gene expression studies in MS using microarrays have identified potential biomarkers of IFN-β response, but none of these has been validated across studies.[2] A major problem with studies on gene expression in this field has been an inability to compare results because of variations in time of blood draw after injection with IFN-β, different doses, routes of administration, preparations of recombinant IFN-β, and

Address for correspondence: Richard M. Ransohoff, M.D., Neuroinflammation Research Center, Department of Neurosciences, Lerner Research Institute, NC30, Cleveland Clinic, 9500 Euclid Avenue, Cleveland, OH 44195. Voice: 216-444-0627; fax: 216-444-7927. ransohr@ccf.org

Cytokine Therapies: Ann. N.Y. Acad. Sci. 1182: 58–68 (2009).
doi: 10.1111/j.1749-6632.2009.05068.x © 2009 New York Academy of Sciences.

variations in MS disease activity, severity, and duration.[3-7]

Biological effects of IFNs are initiated by transcriptional induction of interferon-stimulated genes (ISGs).[8,9] High-density microarrays have been used to identify genes induced by IFNs.[10-12] We selected 162 IFN-β-inducible genes for evaluation, using *ex-vivo* blood samples of treatment-naïve MS patients before and after injection with IFN-β. We developed a customized cDNA macroarray assay for detecting ISG expression which is reproducible, convenient, sensitive, and quantitative.[13]

We propose the hypothesis that the primary molecular responses to IFN-ß injections mediate beneficial and deleterious clinical responses to treatment in MS patients. Variability in clinical response and side effects suggests that there will be differences in ISG expression between individuals. Our hypothesis would be testable only if individuals demonstrated stable ISG signatures over time. Preliminary studies have documented individual variability in the expression of ISGs, but no prior studies have examined the stability of this response.[14,15] We reasoned that relating patterns of ISG induction to therapeutic response could lead to biomarkers for the therapeutic response, and might also provide insight into MS pathogenesis.

Here, we report optimization of a macroarray assay for longitudinal studies of ISG induction in *ex-vivo* blood samples from treatment-naïve MS patients before and after IFN-β injections using a standard dose, route, and IFN-β preparation. We developed a novel bioinformatics methodology for data analysis and compared individual ISG responses at the first injection and after 6 months of treatment. We found marked differences in the number of ISGs induced, in their identity and in the magnitude of induction. However, ISG expression signatures were stable over 6 months of weekly injections for the large majority of patients. Some patients showed identifiable causes for ISG expression inconsistency, including in-

tercurrent viral infection, or neutralizing IFN antibodies. Our results establish conditions for identifying biomarkers of the clinical response to IFN-β in MS. We also propose an overall strategy for monitoring expression signatures in response to transcriptional regulatory therapies for poorly-understood chronic disorders.

## Materials and Methods

### Sample Collection and Patient Information

The study was approved by the Institutional Review Board of the Cleveland Clinic and written informed consent was obtained from all individuals enrolled in the study. Thirty-five patients with relapsing remitting MS (RRMS) or clinically isolated syndromes (CIS, the first clinical episode of RR-MS) who were naïve to treatment were analyzed at the time of their first IFN-ß injection and after 6 months of weekly injections. For ISG analysis, blood (20 ml) was collected directly into PAXgene™ tubes according to manufacturer's instructions 12 h before, and 12 h after an intramuscular injection of 6 million IU of recombinant IFN-β-1a (Avonex) at first injection (baseline) and after 6 months of treatment with IFN-β. Patients had gadolinium-enhanced MRI brain scans at baseline and after 6 months. Expanded Disability Status Scale, Multiple Sclerosis Functional Composite, cell count, differential, and liver enzymes was done at baseline and 6 months. Neutralizing antibody analysis was done at 6 months. The research nurse also collected information on relapses, viral infections, and adverse events known to be associated with IFN-β therapy. Patients were asked to rate the presence and severity of flu-like symptoms, muscle aches, chills, fatigue, headache, and loss of strength on a 11-point Likert-type scale, ranging from 0 (*no side effect at all*) to 10 (*worst you can imagine*) via structured telephone interview 2–3 days after the baseline, 3 month, and 6 month injections.

**TABLE 1.** Patient Characteristics

| | |
|---|---|
| Number of subjects | 35 |
| Females/Males | 23/12 |
| Whites/Blacks | 30/5 |
| Mean age, Male/Female | 34/38 |
| MS type, RR/CIS | 27/8 |
| Mean baseline EDSS (±SD) | 2.0 ± 0.9 |
| Mean baseline MSFC (±SD) | 0.2 ± 0.4 |

RR = Relapsing-Remitting; CIS = Clinically Isolated Syndrome; EDSS = Expanded Disability Status Scale; MSFC = Multiple Sclerosis Function Composite.

Patient characteristics are shown in Table 1. The subjects averaged 37 years of age, 85% were Caucasians, females made up 65% of the group, 78% had RR-MS, and 22% had CIS together with multiple MRI brain lesions.

## RNA Isolation

RNA was extracted *ex-vivo* from blood using PAXgene™ RNA blood extraction kit (PreAnalytix, Switzerland) as per the manufacturer's instructions and concentrated by ethanol precipitation. RNA quality and quantity was assessed by spectrophotometry (absorbance ratios of 280/260 nm) and additional visualization by agarose gel electrophoresis. RNA samples were stored at −80°C.

## Gene Expression using Macroarray

The detailed methodology for cDNA macroarray analysis has been described elsewhere.[13] Genes on the custom array comprised 186 human cDNAs that were primarily selected from the Unigene database. A list of the names of all genes on the macroarray with GenBank accession numbers is shown in Table 2.

Genes on the cDNA macroarray were originally identified from microarray analysis of fibrosarcoma, epithelial or endothelial cell lines treated with IFN-β.[10,11,13] All the genes constituted known ISGs and genes of potential interest and included genes involved in IFN signaling, cytokine production, antiviral, antiproliferative, and immuno-modulatory functions.

The protocol for spotting DNA on the membrane, probe labeling, and hybridization was as reported earlier with local modifications.[13] Five μg of total RNA isolated *ex-vivo* from blood was used for generating radiolabeled cDNA probes by reverse transcription with Superscript II in the presence of [32]PdCTP (Invitrogen, Carlsbad, CA). Residual RNA was hydrolyzed by alkaline treatment at 70°C for 20 min after which cDNA was purified using G50 columns (GE Healthcare, Buckingham shire, UK). Preparation of macroarrays and hybridization of radioactive cDNA were conducted as described previously.[13] Probes were hybridized overnight to macroarray membranes in 10 ml of hybridization buffer, followed by wash with low and high stringency buffers, and exposure to intensifying phosphor screens for two days and scanning by StormImager (Molecular Dynamics, Sunnyvale, CA). Radioactivity bound to the membrane was quantitated and used to calculate induction ratios (IR) of the ISGs as shown below.

To minimize variability, each patient's samples at baseline (0 months) and 6 months were processed in a single batch experiment (total of 4 membranes). A detailed laboratory protocol for the macroarray method is available on request.

## Statistical Analysis

The heatmaps were generated from complete linkage hierarchical cluster analyses. The Euclidean distance metric d used in the cluster analyses is,

$$d = |\mathbf{x} - \mathbf{y}| \sqrt{\sum_{i=1}^{n} |x_i - y_i|^2}.$$

When clustering the subjects, $d_{ij}$ is computed based on the data profiles of all the ISG genes between subjects $i$ and $j$, whereas when clustering the genes, $d_{ij}$ is computed based on the data profiles across all subjects between genes $i$ and $j$. Pearson correlation between baseline and 6 months ISG fold-induction intensity was computed for 35 patients.

**TABLE 2.** List of Genes on the Macroarray

| Gene | Accession No. | Gene | Accession No. | Gene | Accession No. | Gene | Accession No. |
|------|---------------|------|---------------|------|---------------|------|---------------|
| 2–5OAS | NM_002534 | G1P3 | NM_002038 | IP-10 | X02530 | PDK2 | NM_002611 |
| a1-AT | K01396 | Gadd45 | M60974 | IRF4 | U52682 | PGK | V00572 |
| ADAM17 | U69611 | GATA 3 | X58072 | IRF1 | L05072 | PI3K | NM_006219 |
| Adaptin | AF068706 | GBP2 | M55543 | IRF2 | X15949 | PIAS | AF077954 |
| Akt-1 | NM_005163 | Gran B | M17016 | IRF7 | U73036 | PIAS1 | AF077951 |
| Akt-2 | M77198 | HLADP | M83664 | ISG15-L | M13755 | Pig7 | AF010312 |
| APOL3 | AA971543 | HLADRA | J00194 | ISG20 | NM_002201 | PKR | NM_002759 |
| ATF 2 | X15875 | HLAE | X56841 | ISGF3g | M87503 | plectin | U53204 |
| Bad | U66879 | Hou | U32849 | JUN | J04111 | PLSCR1 | AF098642 |
| Bax | U19559 | HPAST | AF00144 | L1CAM | M74387 | PSMB9 | X66401 |
| Bcl-2 | M14745 | Hsf1 | M64673 | L-Selectin | M25280 | Raf | X03484 |
| BST2 | D28137 | Hsp90 | X15183 | MAP2K3 | NM_002756 | RCN1 | D42073 |
| C1-INH | NM_000062 | IDO | NM_002164 | MAP2K4 | L36870 | RGS2 | NM_002923 |
| C1orf29 | NM_006820 | IFI16 | M63838 | MAP3K11 | NM_002419 | RHO GDP | L20688 |
| C1r | NM_001733 | IFI-17 | J04164 | MAP3K14 | NM_003954 | Ribonuc | NM_003141 |
| C1S | J04080 | IFI35 | U72882 | MAP3K3 | U78876 | RIG-1 | AF038963 |
| Caspase 1 | M87507 | IFI44 | D28915 | MAP3K4 | NM_005922 | SERPIN | NM_000295 |
| Caspase 7 | U67319 | IFI-44 | D28915 | MAP3K7 | NM_003188 | Smad1 | U59423 |
| Caspase 9 | U60521 | IFI60 | AF083470 | MAP4K1 | NM_007181 | SNN | NM_003498 |
| CBFA | NM_004349 | IFIT1 | M24594 | MAPK13 | AF004709 | SOCS-1 | N91935 |
| CCR1 | L09230 | IFIT2 | NM_001547 | MAPK7 | NM_002749 | SOCS2 | AF020590 |
| CCR5 | U54994 | IFIT4 | NM_001549 | Met-onco | NM_000245 | SSA1 | NM_003141 |
| CD14 | NM_000591 | IFIT5 | NM_012420 | MIP-1b | NM_002984 | STAT1 | M97935 |
| CD3e | NM_012099 | IFITM2 | NM_006435 | MMP-1 | M13509 | STAT2 | M97934 |
| CEACAM | NM_001712 | IFITM3 | X57352 | MMP-9 | NM_004994 | STAT4 | L78440 |
| c-fos | NM_005252 | IFN-17 | M13755 | MT1H | NM_005951 | STAT5A | L41142 |
| c-myc | L00058 | IFN-9/27 | J04164 | MT1X | NM_005952 | TAP1 | X57522 |
| Collagen | J03464 | IFNAR1 | J03171 | MT2A | NM_005953 | TFEC | NM_012252 |
| COMT | M58525 | IFNAR2 | L42243 | MX1 | M33882 | TGFbR2 | D50683 |
| CREB | NM_004379 | IFNGR1 | J03143 | MX2 | M30818 | TGFbR3 | L07594 |
| CXCL11 | NM_005409 | IFNGR2 | U05875 | NF-IL-6 | X52560 | TIMP-1 | M59906 |
| CXCR4 | AF005058 | IkBa | M69043 | NFkB | M58603 | TNF-a | X01394 |
| CYB56 | NM_007022 | IL15 | U14407 | NMI | Y00664 | TNFAIP6 | NM_007115 |
| Cyp19 | M28420 | IL18 BP | AB019504 | NT5e | X55740 | TOR1B | NM_014506 |
| DDX17 | U59321 | IL1RN | NM_000577 | OASL | NM_003733 | TRAIL | U37518 |
| Def-a3 | NM_005217 | IL2 | NM_000586 | P4HA1 | M24486 | UBE2L6 | NM_004223 |
| Destrin | S65738 | IL2Rg | NM_000206 | p53 | M14694 | USP18 | NM_017414 |
| Elastase 2 | M34379 | IL6 | X04602 | p57Kip2 | U22398 | VegFC | U43142 |
| F-actin | U56637 | IL8Rb | NM_001557 | p70 K | M60724 | Viperin | AF026941 |
| Fas-L | U08137 | iNOS | U20141 | PAI-1 | M16006 | WARS | X62570 |
| FK506 | AF038847 | Int-6 | U62962 | PDGF-a | X06374 | | |
| FLJ20035 | AK000042 | integ-b-6 | NM_000888 | PDK1 | Y15056 | | |

## Results and Discussion

The molecular response to interferon-β in 35 treatment-naïve MS patients was studied at baseline (initial injection) and 6 months with standardized dose, route, and preparation of IFN-β as well as the time elapsed between IFN injection and blood draw. A customized cDNA macroarray assay was used for assessment of ISGs expression signatures.

## Optimization of Macroarray Assay

### Selection of Timing of Phlebotomy before and after Injection with IFN-β

ISGs are subject to differential transcriptional and posttranscriptional control resulting in differences in rates of mRNA accumulation and decay.[16] We selected the 12 h postinjection time point for collection of blood based on an earlier kinetic microarray study which demonstrated that the 12 h time point captured peak induction of the largest number of ISGs involved in the primary response to IFN-β.[4]

### Background Correction and GAPDH Normalization for IR

The IR was defined as the signal from ISG normalized to the GAPDH signal of the postinjection membrane divided by the normalized hybridization signal for the same ISG determined from the preinjection membrane. The induction ratio (IR) of the ISGs was computed using calibrated data as follows: In consideration of the uniformly higher hybridization signal for GAPDH and for empty unspotted background wells on post-IFN membranes, a data imputation rule was applied. For a given membrane, mean plus 2 standard deviations (SDs) of the unspotted wells from the membrane was computed. Any ISGs or GAPDH on the membrane whose intensity values fell below the mean + 2 SD background threshold were replaced with the threshold value. There are four GAPDH triplicate wells on the membrane, and median GAPDH intensity was used in the normalization. Following this calibration, IR of the ISGs was computed as shown below:

IR of ISG

$$= \frac{\text{Gene of interest (post GAPDH (post)}}{\text{Gene of interest (pre GAPDH (pre)}}$$

Where both post- and preinjection values were imputed, the IR ratio was set to 1.

### Calibration Detection Algorithm

ISG cDNAs were spotted on macroarray membrane in triplicate. Spots were occasionally omitted, resulting in outlying data sets (Fig. 1 *left panel* middle spot on the last set of triplicate spots on horizontal row 10; middle spot in the fourth set of triplicate on horizontal row 8). Although membranes were spotted in sets of eight, missing spots could be observed at positions spotted correctly on companion membranes, indicating there was no systematic malfunction of specific pin in the replicator.

The quantitated intensity data are shown in a 3D scatter plot with each coordinate corresponding to one of the three measurements from the triplicate (Fig. 2A). The figure shows that there are two different data patterns. The vast majority of the data locate on the axis that goes from the lower-left corner to the upper-right corner and a small proportion of data locate outside the main data pattern. These two patterns are related to two different sources of measurement variations: one is the biological variation in ISG expression, and the other derives from the measurement process.

To take advantage of the triplicate design to reduce measurement error, the multivariate outlier detection algorithm was applied.[17] Robust distance (RD) was determined using the formula below,

$$RD_i(x_i) = \sqrt{(x_i - \hat{\mu})^T \hat{\Sigma}^{-1}(x_i - \hat{\mu})}$$

where $(\hat{\mu}, \hat{\Sigma})$ are the minimum covariance determinant estimates (MCD) of the location and scatter computed with the FAST-MCD algorithm.[18] Here, RD follows a $\chi^2$ distribution with $df = 3$. The blue dots in Figure 2B represent the data from the ISG that required calibration. For those ISGs, the average signals are calculated only based on the two stable measurements.

Comparison of the measurement error (max–min) versus the average measurement

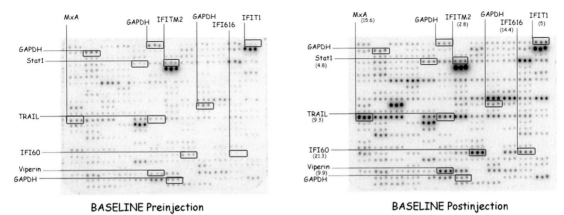

BASELINE Preinjection                    BASELINE Postinjection

**Figure 1.** Digitized image of a macroarray experiment containing selected ISGs. RNA was isolated *ex-vivo* from whole blood of a patient 12 h before and 12 h after IFN-β injection. Each nylon filter was spotted in triplicate with DNA amplified from IMAGE clones representing 186 ISGs. RNA was isolated from whole blood, reverse-transcribed using radiolabel and hybridized to the membranes. All membranes after wash were exposed on StormImager (Molecular Dynamics, Sunnyvale, CA) screen for 48 h and the radioactivity bound to the membrane was quantitated. For the purpose of illustration, 8 induced genes (from a total of 162 genes) and the housekeeping gene glyceraldehyde-3-phosphate dehydrogenase (*GAPDH*) which is spotted on four different locations on the membrane are highlighted. The induction ratios (IR) are shown within parenthesis on the right panel. The left panel shows a preinjection macroarray and on the right is the postinjection macroarray.

of calibrated data (Fig. 2B, right *panel*) and original data (*left panel*) shows clearly that the proposed calibration algorithm is effective in reducing measurement error.

## Assay Precision Supports Defining Gene Induction as an Induction Ratio ≥ 2

Blood was collected from four healthy controls at a 24-h interval and RNA was isolated

**A**

**B**

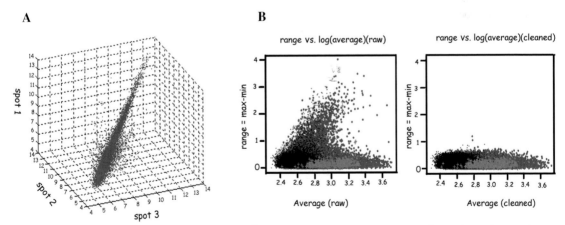

**Figure 2.** Quantitation of intensity data after correction. (**A**) A 3D scatter plot with each coordinate corresponding to one of the three measurements from the triplicate spot for each cDNA on the macroarray. (**B**) Correction of data using the multivariate outlier detection algorithm. The left panel shows the triplicate data spots for all the genes plotted prior to correction and the right panel shows the corrected data after application of the algorithm.

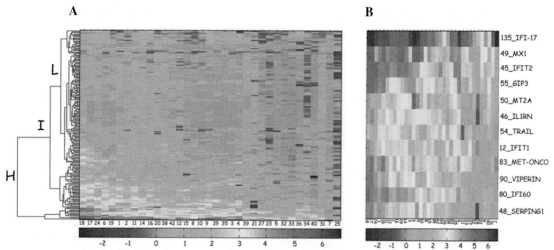

**Figure 3.** ISG expression patterns in MS patients. **(A)** The figure shows a heat map of the ISG fold-induction intensity for 35 patients at baseline. Patients are represented by columns and genes across the rows. The colors range from low induction ratios in blue, to high induction ratios in red. As is evident, the 35 subjects range from high ISG inducers on the left, to low ISG inducers on the right. The heat map illustrates qualitative and quantitative variability in ISG induction across the 35 patients. The ISGs induced in the patients blood after IFN-β injection at baseline can be clustered into three groups as high expression (H), intermediate expression (I), and low expression (L) cluster. **(B)** Expanded view of the high expression cluster shown in the heat map in Figure 3A.

from whole blood. Radiolabeled cDNA probes were generated and hybridized to the genes on macroarray membranes. ISG expression was analyzed after normalization to GAPDH (data not shown). The induction ratio (IR) was defined as the signal from the ISG normalized to the GAPDH signal from the 24-h phlebotomy divided by normalized hybridization signal for the same ISG determined from baseline (0 h) phlebotomy.

The average IR for the four healthy controls clustered around 1 and no ISG showed statistical evidence of IR different from 1. Hence, for the differential expression analysis in MS patients before and after injection with IFN-β, the fold change parameter for significance was set at 2-fold.

## Postinjection ISG Expression Patterns in MS Patients

A representative experiment of a macroarray for ISGs at baseline preinjection and postinjection of IFN-β is shown in Figure 1. For each patient, baseline ISGs (>2 fold induced) were visualized using hierarchical cluster analysis (Fig. 3A). Patients 7 and 25 each of whom had symptoms of viral upper respiratory infection at the baseline time point showed near maximal ISG expression in the preinjection sample

**TABLE 3.** High Expression Cluster of 12 Universally Expressed ISGs in 35 Patients

| Gene | % IR > 2.0 | Mean IR | SD | Min | Max |
|------|-----------|---------|------|------|-------|
| *IFI-17* | 94.3 | 41.0 | 28.1 | 0.1 | 103.0 |
| *MX1* | 97.1 | 17.8 | 12.5 | 1.5 | 52.6 |
| *IFIT2* | 94.3 | 15.2 | 9.8 | 1.5 | 37.4 |
| *G1P3* | 94.3 | 11.6 | 8.8 | 1.4 | 39.8 |
| *IFI60* | 85.7 | 10.7 | 7.9 | 1.1 | 35.8 |
| *Met-onco* | 91.4 | 8.8 | 7.0 | 0.9 | 29.3 |
| *MT2A* | 91.4 | 8.0 | 6.8 | 0.8 | 36.8 |
| *Viperin* | 91.4 | 7.8 | 6.1 | 1.2 | 24.5 |
| *C1-INH* | 80.0 | 7.2 | 5.8 | 0.5 | 23.4 |
| *IL1RN* | 97.1 | 6.8 | 4.0 | 0.6 | 18.8 |
| *IFIT1* | 91.4 | 6.7 | 4.3 | 0.9 | 19.6 |
| *TRAIL* | 88.6 | 6.5 | 3.8 | 0.7 | 14.3 |

IR = Induction Ratio.

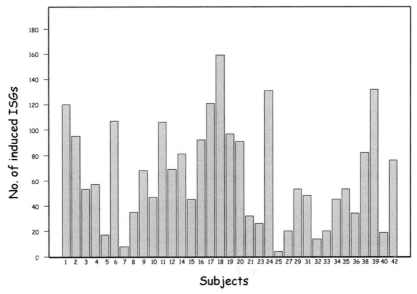

**Figure 4.** Interindividual variation in the number of ISGs induced in MS patients. The histogram shows the number of ISGs with >2-fold induction at baseline for all 35 patients. The subject ID is shown on the x-axis and the number of induced genes is shown on the y-axis.

and did not demonstrate ISG induction at this time point.

The ISGs induced in the patients' blood after IFN-β injection at baseline can be clustered into three groups – high expression, intermediate expression, and low expression cluster. The high expression cluster consisted of 12 universally expressed ISGs which were upregulated

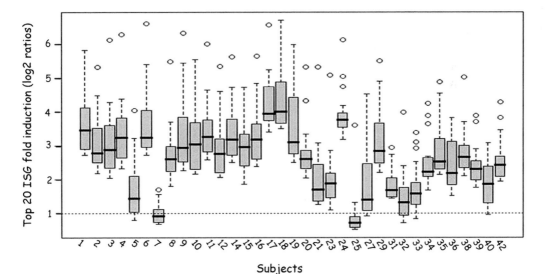

**Figure 5.** Quantitative variability across patients in the magnitude of ISG induction. The figure illustrates the fold-induction of top 20 ISGs for all 35 patients at baseline. This figure shows box and whisker plots for each of the 35 subjects and represents the fold-induction on a log scale for the top 20 genes induced in each subject. All but two subjects (7 and 25) induced numerous genes, but with considerable variability across the subject pool. The figure illustrates the quantitative variability across patients in the magnitude of ISG induction.

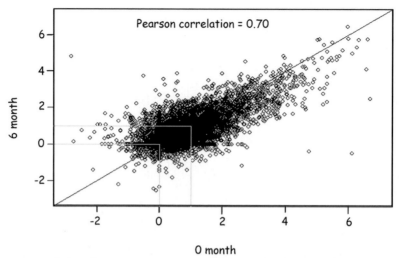

**Figure 6.** Stability of ISG expression. The scatter plot shows the fold-induction for individual genes on the macroarray at baseline (x-axis) and at 6 months (y-axis) for all 35 subjects. The Pearson correlation coefficient was = 0.70.

in almost all patients as shown in Figure 3B and the mean induction ratio averaged from 6- to 41-fold (Table 3). These ISGs included antiviral ISGs like *MX1*, viperin, *IFIT1*, and *IFIT2*; anti-apoptotic genes *G1P3* and met-oncogene; immuno-regulatory genes like *TRAIL, MT2A, C1-INH, IFI-17, IL1RN*, and anti-proliferative gene, *IFI-60*. The intermediate expression cluster comprised 44 ISGs and the rest belonged to the low expression cluster.

We assessed the potential for correlations of the mean of the top 12 ISG ratios at baseline to age, gender, race, disease severity (expanded disability status scale), and type of MS (relapsing-remitting; clinically isolated syndrome). No demographic or disease-related attributes were correlated to the mean fold induction of the top 12 induced ISG expression. Further analysis is in progress to correlate individual gene expression to clinical disease attributes.

### Heterogeneity in IFN-β-induced Gene Expression between Patients at Baseline

The number of induced ISGs (>2 fold) varied among patients (Fig. 4). The number of upregulated ISGs ranged from 4 to 159 with more than 80% of patients showing induction of more than 20 ISGs. Patient 7 and 25 showed the least ISG induction due to high preinjection ISG expression. The number of induced ISGs and the specific genes induced varied among patients (Figs. 3 and 4).

The magnitude of induction of individual ISGs also varied among patients as shown by analysis of the fold-induction of the top 20 ISGs at baseline (Fig. 5).

### Stability of ISG Expression Signatures Over Time

The molecular response to IFN-β injections at 6 months is shown in comparison to baseline by scatter plot in Figure 6. The patients demonstrated stability of ISG expression signature over time, defined as a strong correlation between ISG fold-induction at 6 months compared with baseline. The Pearson correlation coefficients for all the 35 patients was >0.7 (Fig. 6).

### Conclusion

This report provides the first detailed characterization of the primary molecular response to

IFN injections in treatment-naïve MS patients. The molecular signatures of ISG expression detected *ex-vivo* in blood samples from treatment-naïve MS patients after injections with IFN-β at baseline and 6 months varied considerably between individuals. Similar observations were also reported in a recent gene expression study of nine MS patients.[19] We also observed that the ISG response was remarkably stable within individuals over time. The stability of the individual molecular signatures in these patients will enable the identification of molecular biomarkers of the therapeutic response to IFN-β.

During the study, novel methodologies for cDNA macroarray, data analysis, and bioinformatics were developed, and could be useful for studies addressing the identification of molecular biomarkers for therapeutic response, in cases where therapeutic agents regulate mRNA accumulation. Our findings suggest that differential clinical responses among patients will be related to variability in the primary molecular responses to IFN-β injections. Analysis of ISG expression in relation to treatment failure and side effects will be required to address our underlying hypothesis. The results to date suggest that this analysis will be feasible.

## Acknowledgments

We thank Ian Kerr (Cancer Research, UK) and George Stark (Molecular Genetics, Cleveland Clinic, Cleveland) for their valuable advice, and helpful discussions in starting this project. Special thanks to John Barnard, Swathi Chakraborthy and Minnie Chacko at Quantitative Health Sciences, Cleveland Clinic for help in setting up the database, Parianne Fatica, Dianne Ivancic and Cynthia Schwanger at Mellen Center for Multiple Sclerosis, Cleveland Clinic for sample collection.

This research is supported by NIH/NINDS (P50 NS38667, RMR, Program Project PI; Project #4 leader Richard A. Rudick with Richard M. Ransohoff, co-investigator). The procurement of patient samples was supported in part by the National Institutes of Health, National Center for Research Resources, General Clinical Research Center Grant MO1 RR-018390.

## Conflicts of Interest

The authors declare no conflicts of interest.

## References

1. Ann, M.R. & R.A. Rudick. 2006. Drug Insight: interferon treatment in multiple sclerosis. *Nat. Clin. Pract. Neurol.* **2:** 34–44.
2. Goertsches, R. & U.K. Zettl. 2007. MS therapy research applying genome-wide RNA profiling of peripheral blood. *Int. MS J.* **14:** 98–107.
3. Sturzebecher, S., K.P. Wandinger, A. Rosenwald, *et al*. 2003. Expression profiling identifies responder and non-responder phenotypes to interferon-beta in multiple sclerosis. *Brain* **126:** 1419–1429.
4. Weinstock-Guttman, B., D. Badgett, K. Patrick, *et al*. 2003. Genomic effects of IFN-beta in multiple sclerosis patients. *J. Immunol.* **171:** 2694–2702.
5. Wandinger, K.P., C.S. Sturzebecher, B. Bielekova, *et al*. 2001. Complex immunomodulatory effects of interferon-beta in multiple sclerosis include the upregulation of T helper 1-associated marker genes. *Ann. Neurol.* **50:** 349–357.
6. Achiron, A. & M. Gurevich. 2006. Peripheral blood gene expression signature mirrors central nervous system disease: the model of multiple sclerosis. *Autoimmun. Rev.* **5:** 517–522.
7. Santos, R., B. Weinstock-Guttman, M. Tamano-Blanco, *et al*. 2006. Dynamics of interferon-beta modulated mRNA biomarkers in multiple sclerosis patients with anti-interferon-beta neutralizing antibodies. *J. Neuroimmunol.* **176:** 125–133.
8. Stark, G.R., I.M. Kerr, B.R. Williams, *et al*. 1998. How cells respond to interferons. *Annu. Rev. Biochem.* **67:** 227–264.
9. Borden, E.C., G.C. Sen, G. Uze, *et al*. 2007. Interferons at age 50: past, current and future impact on biomedicine. *Nat. Rev. Drug Discov.* **6:** 975–990.
10. Der, S.D., A. Zhou, B.R. Williams, *et al*. 1998. Identification of genes differentially regulated by interferon alpha, beta, or gamma using oligonucleotide arrays. *Proc. Natl. Acad. Sci. USA* **95:** 15623–15628.
11. Miura, A., R. Honma, T. Togashi, *et al*. 2006. Differential responses of normal human coronary artery

endothelial cells against multiple cytokines comparatively assessed by gene expression profiles. *FEBS Lett.* **580:** 6871–6879.

12. Rani, M.R., J. Shrock, S. Appachi, *et al.* 2007. Novel interferon-beta-induced gene expression in peripheral blood cells. *J. Leukoc. Biol.* **82:** 1353–1360.

13. Schlaak, J.F., C.M. Hilkens, A.P. Costa-Pereira, *et al.* 2002. Cell-type and donor-specific transcriptional responses to interferon-alpha. Use of customized gene arrays. *J. Biol. Chem.* **277:** 49428–49437.

14. Whitney, A.R., M. Diehn, S.J. Popper, *et al.* 2003. Individuality and variation in gene expression patterns in human blood. *Proc. Natl. Acad. Sci. USA* **100:** 1896–1901.

15. Hilkens, C.M., J.F. Schlaak, & I.M. Kerr. 2003. Differential responses to interferon-a subtypes in human

Tcells and dendritic cells. *J. Immunol.* **171**(10): 5255–5263.

16. Friedman, R.L., S.P. Manly, M. McMahon, *et al.* 1984. Transcriptional and posttranscriptional regulation of interferon-induced gene expression in human cells. *Cell* **38:** 745–755.

17. Rousseeuw, P.J. & A.M. Leroy. 1987. *Robust Regression and Outlier Detection.* Wiley. New York.

18. Rousseeuw, P.J. & K. Van Driessen. 1999. A fast algorithm for the minimum covariance determinant estimator. *Technometrics* **41:** 212–223.

19. Reder, A.T., S. Velichko, K.D. Yamaguchi, *et al.* 2008. IFN-â1b induces transient and variable gene expression in relapsing-remitting multiple sclerosis patients independent of neutralizing antibodies or changes in IFN receptor RNA expression. *J. Interferon Cytokine Res.* **25:** 317–331.

# Clinical Use of Interferon-γ

## Catriona H.T. Miller,[a] Stephen G. Maher,[b] and Howard A. Young[a]

[a] Center for Cancer Research, Cancer and Inflammation Program, Laboratory of Experimental Immunology, National Cancer Institute-Frederick, Frederick, Maryland, USA

[b] Department of Surgery, Trinity Centre for Health Sciences, Trinity College Dublin, St. James's Hospital, Dublin 8, Ireland

Interferon gamma (IFN-γ), a pleotropic cytokine, has been shown to be important to the function of virtually all immune cells and both innate and adaptive immune responses. In 1986, early clinical trials of this cytokine began to evaluate its therapeutic potential. The initial studies focused on the tolerability and pharmacology of IFN-γ and systematically determined its antitumor and anti-infection activities. In the 20-plus years since those first trials, IFN-γ has been used in a wide variety of clinical indications, which are reviewed in this article.

*Key words:* Interferon-γ; Actimmune; HuZaf

## Introduction

Interferon gamma (IFN-γ), a cytokine with diverse roles in the innate and adaptive immune responses, was discovered in 1965. IFN-γ function has been strongly conserved throughout evolution and across multiple species. T cells, NK cells, and NKT cells are the primary producers of IFN-γ, and it has a myriad of effects in both host defense and immune regulation, including antiviral activity, antimicrobial activity, and antitumor activity. Experimental models in which IFN-γ production has been disrupted have resulted in an increase in autoimmune diseases (reviewed in Ref. 1).

IFN-γ has been shown to be important to the function and maturation of multiple immune cells. IFN-γ is essential for Th1 immune responses and regulates T cell differentiation, activation, expansion, homeostasis, and survival. Killing of intracellular pathogens requires IFN-γ production by T cells. T regulatory cell (Treg) generation and activation requires IFN-γ. IFN-γ stimulates dendritic cells and macrophages to upregulate major histocompatibility complex (MHC) molecules, enhances antigen presentation, and increases expression of costimulatory molecules. IFN-γ stimulated macrophages also produce reactive nitrogen intermediates. NK cells secrete IFN-γ early in host infection, facilitating immune cell recruitment and activation. IFN-γ also activates NK cells, enhancing their cytotoxicity and cell mediated-immune responses. During a viral infection, IFN-γ mediates IgG2a isotype class switching in B cells. IFN-γ recruits neutrophils, stimulates them to upregulate chemokines and adhesion molecules, and triggers rapid superoxide production and respiratory burst (reviewed in Ref. 1). Due to the pleiotrophic effects of IFN-γ on the immune system, it was thought to have great promise as an immunomodulatory drug.

It was not until 1986 that early clinical trials of this cytokine were first conducted.[2] The initial studies were focused on side effects and dose escalation, and then later trials systematically determined its therapeutic potential against cancers and infections. In subsequent years, IFN-γ has been used in a wide variety of clinical

Address for correspondence: Howard A. Young, Laboratory of Experimental Immunology, Cancer and Inflammation Program, National Cancer Institute-Frederick, Building 560 RM31-16, Frederick, MD 21702. Voice: 301-846-5700. younghow@mail.nih.gov

Cytokine Therapies: Ann. N.Y. Acad. Sci. 1182: 69–79 (2009).
doi: 10.1111/j.1749-6632.2009.05069.x © 2009 New York Academy of Sciences.

indications. The most common adverse thera-peutic events occurring with IFN-γ1b therapy are "flu-like," such as fever, headache, chills, myalgia, or fatigue. Other common side effects include rash, injection site erythema or tender-ness, diarrhea and nausea, and leukopenia.[3-8]

IFN-γ clinical trials have been conducted using recombinant derived protein (IFN-γ1b, Actimmune), adenovirus vectors which express IFN-γ cDNA (TG-1041, TG-1042), and neu-tralizing antibodies against IFN-γ (HuZaf and AMG811). Actimmune has been used to treat a wide variety of diseases, including cancer, tuberculosis, hepatitis, chronic granulamotous disease, osteopetrosis, scleroderma, among oth-ers. Adeno-IFN-γ has been used to treat cu-taenous lymphoma and malignant melanoma. HuZaf has been used against autoimmune dis-eases, including rheumatoid arthritis and mul-tiple sclerosis. Promising results have been seen with all agents and are reviewed here.

## Actimmune

The majority of the clinical trials involving IFN-γ have been performed using Actimmune (InterMune) or IFN-γ1b, a genetically engi-neered form of human IFN-γ. The naturally occurring IFN-γ produced by human periph-eral blood leukocytes (PBLs) has the same pri-mary structure as IFN-γ1b.[9,10] However, these two cytokines differ in that: (1) IFN-γ is glyco-sylated and IFN-γ1b is not, (2) IFN-γ1b is 140 amino acids long while IFN-γ has 143 amino acids, and (3) IFN-γ has blocked pyroglutamate residue N-termini, whereas recombinant IFN-γ1b has methionine at the N-terminus.[9-11]

### Cancer

The role of IFN-γ in the host response to cancer has been the subject of numerous re-search studies. Studies have shown that IFN-γ is vital to tumor surveillance by the immune system and a high correlation between IFN-γ production and tumor regression has been seen in immunotherapy. IFN-γ also has direct antitumor effects, as it is anti-angiogenic, in-hibits proliferation, sensitizes tumors cells to apoptosis, upregulates MHC class I and II ex-pression, and stimulates antitumor immune ac-tivity. There have been mixed results as to the efficacy of IFN-γ in the clinical treatment of various cancers.

In a study of recurring superficial transi-tional bladder carcinoma, it was shown that intravesicle instillations of IFN-γ were effective against cancer recurrence.[12] There were signif-icant increases in T cells, NK cells, ICAM-1$^+$ B cells, and HLA-DR$^+$ cell infiltrating the tumor. Strong IFN-γ-induced expression of *HLA-DR* has been correlated with improved prognosis in colorectal cancer patients.[13]

Ovarian cancer is a leading cause of cancer death and is an ideal target for cytokine targeted treatment.[14-16] The presence of intratumoral IFN-γ producing CD3$^+$ T cells is associated with better prognosis.[17] The standard treat-ment is platinum-based chemotherapy. IFN-γ is synergistic with platinum chemotherapeutics *in vitro* at inhibiting ovarian cancer cell pro-liferation and inducing apoptosis.[18] Intraperi-toneal IFN-γ has been shown to achieve anti-tumor responses against ovarian cancer.[19] In a recent randomized phase III trial, administra-tion of subcutaneous IFN-γ and cisplatin to pa-tients improved complete response rates from 56% to 68%[20] and prolonged progression-free survival. Another study showed that IFN-γ with carboplatin and paclitaxel is safe as a first-line treatment of patients with advanced ovarian cancer. Conversely, a phase III clin-ical trial was ended prematurely in 2006. Ovarian cancer and peritoneal carcinoma pa-tients were treated with carboplatin/paclitaxel chemotherapy alone or with IFN-γ 1b sub-cutaneously. In the second interim analysis, it was found that patients treated with IFN-γ and carboplatin/paclitaxel had a significantly shorter survival rate and more adverse events as compared to patients receiving chemotherapy alone. 39.7% of the IFN-γlb and chemother-apy group had died, in contrast to 30.4% of patients in the chemotherapy only group.[21]

IFN-γ is an approved treatment for adult T cell leukemia (ATL) in Japan. There are several reports which claim that intralesional injections of IFN-γ can induce lasting remissions.[22] IFN-γ treatment was first developed and approved in Japan for treatment of Mycosis fungoides (MF).[23]

Several clinical trials are underway using IFN-γ as an adjuvant for vaccine therapies and chemotherapies (NCT00428272, NCT0049-9772, NCT00824733, NCT00004016).

## Tuberculosis

Tuberculosis (TB), a result of infection with *Mycobacterium tuberculosis*, primarily affects the respiratory system, and the emergence of multi-drug resistant strains (MDR-TB) has led to the need for new therapeutic agents. IFN-γ activates alveolar macrophages, which are important in host immunity against *M. tuberculosis*.[24] Condos *et al.* performed a clinical trial examining the effects of an IFN-γ aerosol on MDR-TB. Stabilization or an increase in bodyweight was observed in all IFN-γ treated patients. Furthermore, sputum smears became negative, and a decrease in the mycobacterial burden was seen. Two months after cessation of treatment, a reduction in cavitary lesion sizes was observed in all patients.[25] Another trial found that co-administration of IFN-γ and anti-TB drugs to tuberculosis patients, resulted in increased levels of *STAT1*, *IRF-1*, and *IRF-9* in BAL cells from lung segments. IFN-γ actively stimulated signal transduction and gene expression in alveolar macrophages in TB patients, thus providing a basis for potential use as an adjuvant therapy in this disease.

Other approaches have evaluated IFN-γ as an adjuvant therapy in addition to a chemotherapeutic cocktail. Saurez-Mendez *et al.* found that intramuscular injection of IFN-γ for 6 months as an adjuvant to chemotherapy led to reduced lesion sizes, negative sputum smears and cultures, and increased body mass index.[4]

## *Mycobacterium avium* Complex (MAC)

Atypical mycobacteria infections have been rising, especially among older women. MAC infection leads to progressive chronic pneumonia and lung disease. Atypical mycobacteria survive and proliferate within host macrophages. Treatment of MAC pulmonary infections is difficult because of high drug resistance. IFN–γ has been shown to be a critical cytokine in the resistance of infected macrophages. Thirty-two patients were treated with either intramuscular IFN-γ and chemotherapy or chemotherapy and placebo. The overall response in the IFN-γ group was significantly better than those treated with chemotherapy alone (72.2% vs. 37.5% complete responders). During the study, 35.7% of the control group died compared to 11.1% of the IFN-γ treated group.[26]

## Idiopathic Pulmonary Fibrosis

Currently, there is no FDA approved drug treatment for idiopathic pulmonary fibrosis (IPF). IPF is the most frequent of the idiopathic interstitial pneumonias and has the worst prognosis.[27] IPF is a chronic condition characterized by progressive scarring, loss of lung function, progressive limitation, and eventual death.[28] Alveolar epithelial cells release fibrogenic cytokines, such as TGF-β, PDGF, TNF-α, IL-1, insulin-like growth factor-1, and basic fibroblast factor, in response to injury. The release of these cytokines causes fibroblast proliferation, migration to the lung, and fibroblast differentiation.[29,30] Traditional therapies have been ineffective, and new agents are required to halt the progression of disease. IFN-γ therapy of IPF has been explored, as IPF is characterized by an IFN-γ deficit. It was hypothesized that treatment with IFN-γ might halt the progression of IPF.

Unfortunately, the results from IPF studies with IFN-γ treatment are mixed. In an initial randomized clinical trial[31] of IPF patients, IFN-γ treatment in combination with prednisolone resulted in increased total lung capacity and increased resting and maximal exertion values of partial pressure of arterial oxygen, compared to prednisolone alone. In a retrospective study of qualified IPF patients, IFN-$\gamma$ had beneficial effects on forced vital capacity and single breath diffusing capacity for $CO_2$.[28] Furthermore, these effects were most pronounced in patients with advanced disease.

Conversely, in a large (330 patients) one year placebo-controlled clinical trial, subcutaneous IFN-$\gamma$ administration did not affect progression-free survival or pulmonary function.[32] The time to death or disease progression was not significantly altered. However, subgroup analyses, which are at best hypothesis generating, showed a possible survival benefit for patients with mild-to-moderate impairment.[33]

The results of a double-blind clinical trial of the molecular effects of subcutaneous IFN-$\gamma$1b in IPF patients were published in 2004.[34] After IFN-$\gamma$1b treatment, expression of the immunomodulatory chemoattractant CXCL11 was increased in bronchoalveolar lavage fluid (BALF) and plasma. Levels of neutrophil activating CXCL5, PDGFA (platelet derived growth factor A), and type 1 procollagen were lower in BALF. Gene expression studies showed increases in *CXCL11*, type III procollagen, and *PDGFB* in transbronchial biopsy samples. A decrease in elastin was also seen. These changes suggested that IFN-$\gamma$ 1b would be an effective treatment for IPF via multiple pathways.[34]

In March of 2007, the INSPIRE trial of Actimmune for treating IPF (NCT00075998) was forced to end prematurely. INSPIRE was a randomized, double-blind, placebo-controlled Phase 3 study designed to evaluate the safety and efficacy of Actimmune in IPF patients with mild to moderate impairment in lung function. The primary endpoint was survival time. An interim analysis showed that patients who received Actimmune did not benefit. Approximately 14.5% of patients treated with Actimmune died compared to 12.7% of placebo treated patients (NCT00075998).

## Cystic Fibrosis

Cystic fibrosis (CF) is an inherited disorder caused by a mutation in a chloride channel, the cystic fibrosis transmembrane conductance regulator gene. CF is characterized by chronic endobronchial infection and inflammation, destruction of lung tissue, and eventual respiratory failure in 90% of patients. CF patients have inefficient pulmonary clearance of thick secretions and defects in production of nitric oxide resulting in chronic bacterial infections with a thick bacterial biofilm. Bacterial biofilms resist opsonins, phagocytes, antibiotics, and cause chronic neutrophil inflammation. Production of elastase by neutrophils contributes to lung damage. Due to the ability of IFN-$\gamma$ to activate macrophages, correct deficiencies in NO production *in vitro*, and inhibit the proliferation of Th2 clones, and based on clinical data showing IFN-$\gamma$ deficiencies in PBMCs from cystic fibrosis patients,[35,36] it was thought to have great potential as a treatment for cystic fibrosis. Sixty-six cystic fibrosis patients received 50 $\mu$g–1000 $\mu$g aerosolized IFN-$\gamma$1b or placebo three times a week for 12 weeks.[37] No statistically significant differences were seen in the primary endpoints of the trial, $FEV_1$ (forced expiratory volume in 1 s) and sputum bacterial density over the entire study. At 4 weeks, a slight, but significant reduction was seen in the $FEV_1$ of the 1000 $\mu$g IFN-$\gamma$ treated group versus placebo, and a significant reduction was also seen in bacterial density between the 1000 $\mu$g IFN-$\gamma$ treated group versus placebo. No significant statistical differences were seen in levels of neutrophils, IL-8, elastase, myeloperoxidase, or DNA in the sputum of either IFN-$\gamma$ or placebo treated patients.

## Hepatitis

Liver cirrhosis and hepatocellular carcinoma arise primarily as a result of chronic hepatitis infection.[5,38] The current regimen for the treatment of chronic hepatitis B (HBV) and C (HCV) is pegylated IFN-$\alpha$ (peginterferon-$\alpha$)

and ribavirin with a response rate of only 50–60%.[6] In light of the large number of non-responders, IFN-γ has been evaluated as a potential alternative treatment. In chronic HBV trials, IFN-γ1b alone was not found to have any significant impact on viral infection but did modulate the immune system.[7] Treatment of HCV with IFN-γ1b has also proven generally unsuccessful[8,39] but pretreatment with IFN-γ prior to IFN-α treatment resulted in enhanced immunologic activity in HCV patients. The enhanced immunological activity is speculated to enhance IFN-α-mediated viral clearance.[40]

Fibrosis accounts for the majority of the complications associated with chronic hepatitis. IFN-γ has been shown to have antifibrotic effects and is efficacious against fibrosis in HBV patients.[41] In a recent study, the antifibrotic activity of IFN-γ1b in HCV patients was examined.[8] Although no overall reduction in fibrosis was seen, select patients had significant fibrosis reductions.

## Chronic Granulomatous Disease

In 1991, IFN-γ1b was FDA approved for the treatment of chronic granulomatous disease (CGD). CGD is an inherited disorder of leukocyte function, caused by a defect or the absence of reduced nicotinamide adenine dinucleotide phosphate oxidase. This enzyme is essential for microbiocidal activity and superoxide generation in phagocytes. Consequently, CGD patients suffer recurrent life-threatening bacterial and fungal infections. As IFN-γ is known to enhance the respiratory burst of human phagocytic cells, it was hoped that IFN-γ would reverse the immunological defects observed in CGD patients. In clinical trials, IFN-γ1b reduced the frequency and severity of serious infections in CGD patients.[11]

Marciano *et al.* examined the long-term effects of IFN-γ1b administration to CGD patients.[42] 76 patients were enrolled in an uncontrolled study to assess long-term safety and efficacy of IFN-γ1b therapy. Patients were followed for up to 9 years and received IFN-γ1b subcutaneously thrice weekly. This study concluded that IFN-γ prophylaxis for CGD appears to be effective and well tolerated over a prolonged period of time. Actimmune product literature claims that treatment with their product leads to a 67% reduction in the relative risk of serious infections, 53% fewer primary infections, 64% fewer infections overall, and 67% fewer inpatient hospital days.

## Osteopetrosis

Congenital osteopetrosis (OP) is a rare osteosclerotic bone disease caused by a defect in osteoclast function and bone resorption.[43] Severe, malignant OP is characterized by an overgrowth of body structures which results in infection, anemia, thrombocytopenia, blindness, deafness, and ultimately early death.[44–46] The granular leukocytes of OP patients are defective in superoxide production,[47,48] resulting in frequent, severe infections. As IFN-γ1b reduces infection in CGD patients by increasing neutrophil superoxide production, Key *et al.* hypothesized that IFN-γ1b might stimulate osteoclasts in a similar manner.[49] IFN-γ1b administration to OP patients significantly increased osteoclastic bone resorption, increased superoxide production in PBLs, and reduced infection.[50] *In vitro* studies with OP patient blood cultures, have shown that IFN-γ1b enhances osteoclast generation and normalizes superoxide production.[45] Presently, the only effective treatment for osteopetrosis is hematopoietic stem cell transplant (HSCT), which has high treatment related morbidity and mortality.[51]

In 2000, the FDA approved Actimmune for delaying the time to disease progression in patients with severe malignant OP. In a 1999 phase III clinical trial, 15 osteopetrosis patients received either Actimmune or control vitamin D. The length of time to disease progression was significantly delayed in patients treated with Actimmune (165 days) compared to patients treated with the control (65 days). Evidence of increased bone resorption, enhanced bone marrow activity, and a reduction in serious infections was observed.[50]

## Scleroderma

Scleroderma is a connective tissue disease that affects multiple organ systems, including skin, heart, lungs, and kidneys.[52] The mechanisms of fibrosis in scleroderma are not fully understood. It is known that soluble mediators, such as TGF-$\beta$, PDGF, IL-4, IL-6, and TNF-$\alpha$, can affect the behavior of fibroblast growth, proliferation, collagen synthesis, and chemotaxis.[52–54] IFN-$\gamma$ has been used in the treatment of scleroderma because of its antifibrotic activity, its ability to reduce collagen production *in vitro*, and to inhibit fibroblast cell proliferation. In most clinical trials, either subcutaneous or intramuscular administration of IFN-$\gamma$ to scleroderma patients has resulted in modest improvements.[55–58]

## Invasive Fungal Infections/Immunosuppressed Patients

Invasive fungal infections are an increasing problem, especially in immunocompromised patients, such as leukemia patients, HIV patients, and transplant patients. The most common infectious agents are candida and aspergillus. There are several new drugs available to treat fungal infection, including triazoles and immunotherapeutics, such as colony stimulating factors, granulocyte transfusions and IFN-$\gamma$. GM-CSF and IFN-$\gamma$ are given in combination to treat patients with serious refractory fungal infections and non-neutropenic infections. It has been demonstrated that IFN-$\gamma$ increases the anti-fungal activity of macrophages and neutrophils.[26] (ISRCTN70900209)

*Cryptococcus neoformans* is responsible for the most common central nervous system infection in HIV patients, acute cryptococcal meningitis, and is the most common cause of fungal meningitis worldwide. In a double-blind clinical trial, patients received either 100 or 200 $\mu$g of IFN-$\gamma$1b or placebo in addition to standard antifungal therapy. (NCT00012467) Among 75 patients, 13% of placebo patients, 36% of 100 $\mu$g IFN-$\gamma$1b, and 32% of 200 $\mu$g IFN-$\gamma$1b had fungus clean cerebrospinal fluid cultures after 2 weeks.[59]

The efficacy of IFN-$\gamma$1b to reduce opportunistic infections in advanced HIV was tested in a 12-month double-blind phase III trial of HIV patients on antiretroviral drugs. Eighty-four patients were treated with either IFN-$\gamma$ or placebo subcutaneously for 48 weeks. Patients on placebo had an average of 3.45 opportunistic infections in the first 48 weeks, while patients with IFN-$\gamma$1b therapy had an average of 1.71. Three-year survival in the IFN-$\gamma$ arm was 28% compared to 18% in the placebo group, although this difference was not statistically significant. IFN-$\gamma$1b treatment was especially effective against candida, herpes, and cytomegalovirus infections.[60]

## Adeno-IFN-$\gamma$

### Cutaneous B and T Cell Lymphomas

Primary cutaneous lymphomas (CL) are characterized by an accumulation of clonal T or B lymphocytes in the skin. Typically, CLs are chronic indolent diseases. TG-1042, or adeno-IFN-$\gamma$, was investigated in a phase I trial, where patients with advanced primary cutaneous B or T CL were repeatedly injected intratumorally with TG-1042.[61,62] TG-1042 is a nonreplicating adenovirus vector containing a human IFN-$\gamma$ cDNA insert. Five of nine treated patients had local clinical responses. Three patients had complete responses with the clearance of non-injected skin lesions. Two patients had partial responses. IFN-$\gamma$ from TG-1042 message was detected in injected lesions in seven patients after the first treatment cycle and remained detectable for several cycles. Humoral antitumor immune responses were also detected. Adeno-IFN-$\gamma$ is now being tested in a phase II trial (NCT00394693) against chronic BCL.[61]

## Malignant Melanoma

A phase I clinical trial was conducted in 2003 using TG-1041, a recombinant adeno-IFN-γ vector to treat malignant melanoma. TG-1041 is a replication deficient adenovirus with the cDNA for IFN-γ inserted in the E1 region of the viral genome. Patients were given three intratumoral injections of adeno-IFN-γ. Out of 11 treated patients, no complete or partial responses were seen. However five patients had minor decreases in injected tumor nodules, eight patients had local inflammation, one patient had significant necrosis of the injected nodule, one patient had inflammation of distant nodules, and one patient had disease stabilization.[63]

## Anti-Interferon-γ Antibodies (HuZaf and AMG811)

Fontolizumab (HuZAF) is a humanized monoclonal antibody that binds IFN-γ and inhibits expression of IFN-γ regulated genes. Fontolizumab is being explored for the treatment of autoimmune diseases, such as Crohn's disease, lupus, rheumatoid a arthritis, and multiple sclerosis. Adverse side effects are generally mild and rare, and include abdominal pain, vomiting, headache, nausea, arthalgia, asthenia, and cough.[64,65] AMG811, a fully human monoclonal antibody that binds and neutralizes IFN-γ, is being evaluated by Amgen.

## Multiple Sclerosis

Multiple sclerosis (MS) is an autoimmune disorder where the immune system attacks the myelin sheath of the central nervous system. Disturbances in cytokine synthesis, particularly IFN-γ, play critical roles in the initiation and prolongation of MS.[66] IFN-β is currently used in the treatment of MS and may work via inhibition of IFN-γ-mediated immune activation. In a randomized study of patients with progressive MS, those patients who received a short course of neutralizing IFN-γ antibody had a significant delay in disability progression.[65] MRIs showed decreased numbers of active lesions. The cytokine profile produced by activated blood cells from treated patients changed, with decreased IL-1β, TNF-α, and IFN-γ and increased TGF-β. These data indicate that neutralizing IFN-γ may be a new treatment option for the management of progressive MS.

## Crohn's Disease

Recent studies have examined the safety and efficacy of HuZAF in the treatment of moderate to severe Crohn's disease (CD).[67] Crohn's disease is a chronic inflammatory disease of the gastrointestinal tract. IFN-γ has been implicated in the inflammation observed in CD and increased levels are found in the mucosa.[64] In several models of experimental colitis, increased mucosal IFN-γ levels were detected. In a randomized double-blind phase II clinical trial, 42 patients received one dose, and 91 patients received two doses of HuZAF.[67] There was no difference in response between HuZAF and placebo groups after a single dose. In contrast, those patients receiving two doses of fontolizumab had a doubled response rate at day 56 compared to placebo controls. Given the long half-life (18 days) and low immunogenicity, this study concluded that treatment of active CD with anti-IFN-γ antibody warrants further investigation.[67] Based on the results with Crohn's disease, HuZAF is being considered as a treatment for pediatric inflammatory bowel disease.

## Rheumatoid Arthritis

Rheumatoid arthritis (RA) is a chronic inflammatory disease characterized by a predominant Th1-associated autoimmunity. Early clinical studies tested IFN-γ as a therapeutic for rheumatoid arthrititis. In these studies, 54 patients were treated with IFN-γ, while 51 received placebo in a double-blind study.[68] While greater improvement was seen in the IFN-γ treated group, it was not statistiscally significant. Recently, antibodies to IFN-γ have been found to be beneficial in the treatment of RA, in a randomized double-blind trial.[69] Thirty patients with active RA received intramuscular injections of either anti-IFN-γ, anti-TNF-α, or placebo. Based on a physical examination, nine patients receiving anti-TNF-α, seven receiving anti-IFN-γ and two receiving placebo, appeared to have an improvement of their condition.

## Lupus

Systemic lupus erythematosus (SLE) is an autoimmune disorder characterized by self-reactive antibodies, often against nucleic acids, which form immune complexes and collect in body organs and joints. Because SLE patients have been shown to have increased levels of serum IFN-γ,[70] IFN-γ has been shown to exacerbate SLE,[71] and in mouse models, IFN-γ receptor is required for development of SLE.[72] Amgen has recently begun a phase Ib clinical trial, using AMG-811, a fully human monoclonal antibody that binds IFN-γ to treat SLE (NCT00818948). This trial will be focused on safety and pharmokinetics.

## Summary

IFN-γ has proven to be a key immunoregulatory molecule whose effects on immune system development, maturation and function is widespread, affecting a myriad of cell types. While IFN-γ affects numerous disease processes, the clinical applications of this important molecule are currently limited. Given the critical importance of IFN-γ in immunity, clinical use of IFN-γ will depend upon a more precise understanding of its basic biology and localized effects in order to better define how to use this molecule in the context of the disease setting.

## Acknowledgment

This project has been funded in whole or in part with federal funds from the National Cancer Institute, National Institutes of Health, under Contract No. HHSN261200800001E. The content of this publication does not necessarily reflect the views or policies of the Department of Health and Human Services, nor does mention of trade names, commercial products, or organizations imply endorsement by the U.S. Government.

## Note

Given the large number of publications on IFN- γ, it is inevitable that some authors may believe that their work was not properly cited or included in this review. The authors of this review apologize to those investigators whose work was not cited as any exclusion of relevant work was not intentional.

## Conflicts of Interest

The authors declare no conflicts of interest.

## References

1. Young, H.A., A.L. Romero-Weaver, R. Savan, *et al.* 2007. Interferon-gamma. *Interferon-γ in Class II Cytokines*. A. Zdanov, Ed.: 51–106. Research Signpost, Kerala, India.
2. Jaffe, H.S. & S.A. Sherwin. 1986. The early clinical trials of recombinant human interferon-gamma. *Interferons as Cell Growth Inhibitors and Antitumor Factors*. R. Freedman, T. Merigan & T. Sreevalson, Eds.: 37–46. Alan R. Liss. New York.
3. Condos, R., F.P. Hull, N.W. Schluger, *et al.* 2004. Regional deposition of aerosolized interferon-gamma in pulmonary tuberculosis. *Chest* **125:** 2146–2155.
4. Suarez-Mendez, R., I. Garcia-Garcia, N. Fernandez-Olivera, *et al.* 2004. Adjuvant interferon gamma in patients with drug-resistant pulmonary tuberculosis: A pilot study. *BMC Infect. Dis.* **4:** 44.
5. Soza, A., T. Heller, M. Ghany, *et al.* 2005. Pilot study of interferon gamma for chronic hepatitis C. *J. Hepatol.* **43:** 67–71.
6. Di Bisceglie, A.M. & J.H. Hoofnagle. 2002. Optimal therapy of hepatitis C. *Hepatology* **36:** S121–S127.
7. Lau, J.Y., C.L. Lai, P.C. Wu, *et al.* 1991. A randomised controlled trial of recombinant interferon-gamma in Chinese patients with chronic hepatitis B virus infection. *J. Med. Virol.* **34:** 184–187.

8. Muir, A.J., P.B. Sylvestre & D.C. Rockey. 2006. Interferon gamma-1b for the treatment of fibrosis in chronic hepatitis C infection. *J. Viral Hepat.* **13:** 322–328.

9. Rinderknecht, E., B.H. O'Connor & H. Rodriguez. 1984. Natural human interferon-gamma. Complete amino acid sequence and determination of sites of glycosylation. *J. Biol. Chem.* **259:** 6790–6797.

10. Rinderknecht, E. & L.E. Burton. 1985. Biochemical characterization of natural and recombinant IFN-gamma. *The Biology of the Interferon System.* H. Kirchner & H. Schellekens, Eds.: 397–402. Elsevier. Amsterdam.

11. Czarniecki, C.W. & G. Sonnenfeld. 2006. Clinical applications of interferon-gamma. *The Interferons: Characterization and Application.* A. Meager, Ed.: 309–336. Wiley-VCH. Weinheim.

12. Giannopoulos, A., C. Constantinides, E. Fokaeas, *et al.* 2003. The immunomodulating effect of interferon-gamma intravesical instillations in preventing bladder cancer recurrence. *Clin. Cancer Res.* **9:** 5550–5558.

13. Matsushita, K., T. Takenouchi, H. Shimada, *et al.* 2006. Strong HLA-DR antigen expression on cancer cells relates to better prognosis of colorectal cancer patients: Possible involvement of c-myc suppression by interferon-gamma in situ. *Cancer Sci.* **97:** 57–63.

14. Chen, C.K., M.Y. Wu, K.H. Chao, *et al.* 1999. T lymphocytes and cytokine production in ascitic fluid of ovarian malignancies. *J. Formos. Med. Assoc.* **98:** 24–30.

15. Punnonen, R., K. Teisala, T. Kuoppala, *et al.* 1998. Cytokine production profiles in the peritoneal fluids of patients with malignant or benign gynecologic tumors. *Cancer* **83:** 788–796.

16. Wall, L., F. Burke, J.F. Smyth & F. Balkwill. 2003. The anti-proliferative activity of interferon-gamma on ovarian cancer: in vitro and in vivo. *Gynecol. Oncol.* **88:** S149–S151.

17. Marth, C., H. Fiegl, A.G. Zeimet, *et al.* 2004. Interferon-gamma expression is an independent prognostic factor in ovarian cancer. *Am. J. Obstet. Gynecol.* **191:** 1598–1605.

18. Marth, C., G.H. Windbichler, H. Hausmaninger, *et al.* 2006. Interferon-gamma in combination with carboplatin and paclitaxel as a safe and effective first-line treatment option for advanced ovarian cancer: results of a phase I/II study. *Int. J. Gynecol. Cancer* **16:** 1522–1528.

19. Pujade-Lauraine, E., J.P. Guastalla, N. Colombo, *et al.* 1996. Intraperitoneal recombinant interferon gamma in ovarian cancer patients with residual disease at second-look laparotomy. *J. Clin. Oncol.* **14:** 343–350.

20. Windbichler, G.H., H. Hausmaninger, W. Stumvoll, *et al.* 2000. Interferon-gamma in the first-line therapy of ovarian cancer: a randomized phase III trial. *Br. J. Cancer* **82:** 1138–1144.

21. Alberts, D.S., C. Marth, R.D. Alvarez, *et al.* 2008. Randomized phase 3 trial of interferon gamma-1b plus standard carboplatin/paclitaxel versus carboplatin/paclitaxel alone for first-line treatment of advanced ovarian and primary peritoneal carcinomas: Results from a prospectively designed analysis of progression-free survival. *Gynecol. Oncol.* **109:** 174–181.

22. Tamura, K., S. Makino, Y. Araki, *et al.* 2006. Recombinant interferon beta and gamma in the treatment of adult T-cell leukemia. *Cancer* **59:** 1059–1062.

23. Nagatani, T., H. Okazawa, N. Inomata, *et al.* 2004. PUVA and interferon-gamma combination therapy for plaque stage mycosis fungoides. *Nishi Nihon Hifuka* **66:** 274–279.

24. Condos, R., B. Raju, A. Canova, *et al.* 2003. Recombinant gamma interferon stimulates signal transduction and gene expression in alveolar macrophages in vitro and in tuberculosis patients. *Infect. Immun.* **71:** 2058–2064.

25. Condos, R., W.N. Rom & N.W. Schluger. 1997. Treatment of multidrug-resistant pulmonary tuberculosis with interferon-gamma via aerosol. *Lancet* **349:** 1513–1515.

26. Milanes-Virelles, M., I. Garcia-Garcia, Y. Santos-Herrera, *et al.* 2008. Adjuvant interferon gamma in patients with pulmonary atypical Mycobacteriosis: A randomized, double-blind, placebo-controlled study. *BMC Infect. Dis.* **8:** 17.

27. Katzenstein, A.L. & J.L. Myers. 1998. Idiopathic pulmonary fibrosis: clinical relevance of pathologic classification. *Am. J. Respir. Crit. Care Med.* **157:** 1301–1315.

28. Nathan, S.D., S.D. Barnett, B. Moran, *et al.* 2004. Interferon gamma-1b as therapy for idiopathic pulmonary fibrosis. An intrapatient analysis. *Respiration* **71:** 77–82.

29. Kahlil, N. & R. O'Connor. 2004. Idiopathic pulmonary fibrosis: current understanding of the pathogenesis and the status of treatment. *CMAJ* **171:** 153–160.

30. Selman, M., V.J. Thannickal, A. Pardo, *et al.* 2004. Idiopathic pulmonary fibrosis: pathogenesis and therapeutic approaches. *Drugs* **64:** 405–430.

31. Ziesche, R., E. Hofbauer, K. Wittmann, *et al.* 1999. A preliminary study of long-term treatment with interferon gamma-1b and low-dose prednisolone in patients with idiopathic pulmonary fibrosis. *N. Engl. J. Med.* **341:** 1264–1269.

32. Raghu, G., K.K. Brown, W.Z. Bradford, *et al.* 2004. A placebo-controlled trial of interferon gamma-1b in

patients with idiopathic pulmonary fibrosis. *N. Engl. J. Med.* **350:** 125–133.

33. Shah, N.R., P. Noble, R.M. Jackson, *et al.* 2005. A critical assessment of treatment options for idiopathic pulmonary fibrosis. *Sarcoidosis. Vasc. Diffuse. Lung Dis.* **22:** 167–174.

34. Strieter, R.M., K.M. Starko, R.I. Enelow, *et al.* & the other members of the Idiopathic Pulmonary Fibrosis Biomarkers Study Group. 2004. Effects of interferon-{gamma} 1b on biomarker expression in patients with idiopathic pulmonary fibrosis. *Am. J. Respir. Crit. Care Med.* **170:** 133–140.

35. Knutsen, A.P., P.S. Hutchinson, G.M. Albers, *et al.* 2003. Increased sensitivity to IL-4 in cystic fibrosis patients with allergic bronchopulmonary aspergillosis allergy. *Allergy* **59:** 81–87.

36. Moss, R.B., Y.P. Hsu & L. Olds. 2009. Cytokine dysregulation in activated cystic fibrosis (CF) peripheral lymphocytes. *Clin. Exp. Immunol.* **120:** 518–525.

37. Moss, R.B., N. Mayer-Hamblett, J. Wagener, *et al.* 2004. Randomized, double-blind, placebo-controlled, dose-escalating study of aerosolized interferon gamma-1b in patients with mild to moderate cystic fibrosis lung disease. *Pediatr. Pulmonol.* **39:** 209–218.

38. Brown, R.S., Jr. & P.J. Gaglio. 2003. Scope of worldwide hepatitis C problem. *Liver Transpl.* **9:** S10–S13.

39. Saez-Royuela, F., J.C. Porres, A. Moreno, *et al.* 1991. High doses of recombinant alpha-interferon or gamma-interferon for chronic hepatitis C: A randomized, controlled trial. *Hepatology* **13:** 327–331.

40. Katayama, K., A. Kasahara, Y. Sasaki, *et al.* 2001. Immunological response to interferon-gamma priming prior to interferon-alpha treatment in refractory chronic hepatitis C in relation to viral clearance. *J. Viral Hepat.* **8:** 180–185.

41. Weng, H.L., W.M. Cai & R.H. Liu. 2001. Animal experiment and clinical study of effect of gamma-interferon on hepatic fibrosis. *World J. Gastroenterol.* **7:** 42–48.

42. Marciano, B.E., R. Wesley, E.S. De Carlo, *et al.* 2004. Long-term interferon-gamma therapy for patients with chronic granulomatous disease. *Clin. Infect. Dis.* **39:** 692–699.

43. Blin-Wakkach, C., A. Wakkach, P.M. Sexton, *et al.* 2004. Hematological defects in the oc/oc mouse, a model of infantile malignant osteopetrosis. *Leukemia* **18:** 1505–1511.

44. Key, L.L., Jr., W.C. Wolf, C.M. Gundberg & W.L. Ries. 1994. Superoxide and bone resorption. *Bone* **15:** 431–436.

45. Madyastha, P.R., S. Yang, W.L. Ries & L.L. Key, Jr. 2000. IFN-gamma enhances osteoclast generation in cultures of peripheral blood from osteopetrotic patients and normalizes superoxide production. *J. Interferon Cytokine Res.* **20:** 645–652.

46. Shapiro, F., M.J. Glimcher, M.E. Holtrop, *et al.* 1980. Human osteopetrosis: A histological, ultrastructural, and biochemical study. *J. Bone Joint Surg. Am.* **62:** 384–399.

47. Beard, C.J., L. Key, P.E. Newburger, *et al.* 1986. Neutrophil defect associated with malignant infantile osteopetrosis. *J. Lab Clin. Med.* **108:** 498–505.

48. Reeves, J.D., C.S. August, J.R. Humbert & W.L. Weston. 1979. Host defense in infantile osteopetrosis. *Pediatrics* **64:** 202–206.

49. Key, L.L., Jr., W.L. Ries, R.M. Rodriguiz & H.C. Hatcher. 1992. Recombinant human interferon gamma therapy for osteopetrosis. *J. Pediatr.* **121:** 119–124.

50. Key, L.L., Jr., R.M. Rodriguiz, S.M. Willi, *et al.* 1995. Long-term treatment of osteopetrosis with recombinant human interferon gamma. *N. Engl. J. Med.* **332:** 1594–1599.

51. Stark, Z. & R. Savarirayan. 2009. Osteopetrosis. *Orphanet J. Rare Dis.* **4:** 5.

52. Sapadin, A.N. & R. Fleischmajer. 2002. Treatment of scleroderma. *Arch. Dermatol.* **138:** 99–105.

53. Fleischmajer, R., J.S. Perlish, T. Krieg & R. Timpl. 1981. Variability in collagen and fibronectin synthesis by scleroderma fibroblasts in primary culture. *J. Invest. Dermatol.* **76:** 400–403.

54. Jimenez, S.A., E. Hitraya & J. Varga. 1996. Pathogenesis of scleroderma. Collagen. *Rheum. Dis. Clin. North Am.* **22:** 647–674.

55. Grassegger, A., G. Schuler, G. Hessenberger, *et al.* 1998. Interferon-gamma in the treatment of systemic sclerosis: a randomized controlled multicentre trial. *Br. J. Dermatol.* **139:** 639–648.

56. Hein, R., J. Behr, M. Hundgen, *et al.* 1992. Treatment of systemic sclerosis with gamma-interferon. *Br. J. Dermatol.* **126:** 496–501.

57. Hunzelmann, N., S. Anders, G. Fierlbeck, *et al.* 1997. Systemic scleroderma. Multicenter trial of 1 year of treatment with recombinant interferon gamma. *Arch. Dermatol.* **133:** 609–613.

58. Polisson, R.P., G.S. Gilkeson, E.H. Pyun, *et al.* 1996. A multicenter trial of recombinant human interferon gamma in patients with systemic sclerosis: Effects on cutaneous fibrosis and interleukin 2 receptor levels. *J. Rheumatol.* **23:** 654–658.

59. Pappas, P., B. Bustamante, E. Ticona, *et al.* 2004. Recombinant interferon-γ1b as adjunctive therapy for AIDS-related acute cryptococcal meningitis. *J. Infect. Dis.* **189:** 2185–2191.

60. Riddell, L.A., A.J. Pinching, S. Hill, *et al.* 2001. A phase III study of recombinant human interferon gamma to prevent opportunistic infections in

advanced HIV disease. *AIDS Res. Hum. Retroviruses* **17:** 789–797.

61. Dummer, R. 2005. Emerging drugs in cutaneous T-cell lymphomas. *Expert Opin. Emerg. Drugs* **10:** 381–392.

62. Dummer, R., J.C. Hassel, F. Fellenberg, *et al.* 2004. Adenovirus-mediated intralesional interferon-{gamma} gene transfer induces tumor regressions in cutaneous lymphomas. *Blood* **104:** 1631–1638.

63. Khorana, A.A., J.D. Rosenblatt, D.M. Sahasrabudhe, *et al.* 2003. A phase I trial of immunotherapy with intratumoral adenovirus-interferon-gamma (TG1041) in patients with malignant melanoma. *Cancer Gene Ther.* **10:** 251–259.

64. Fuss, I.J., M. Neurath, M. Boirivant, *et al.* 1996. Disparate CD4+ lamina propria (LP) lymphokine secretion profiles in inflammatory bowel disease. Crohn's disease LP cells manifest increased secretion of IFN-gamma, whereas ulcerative colitis LP cells manifest increased secretion of IL-5. *J. Immunol.* **157:** 1261–1270.

65. Skurkovich, S., A. Boiko, I. Beliaeva, *et al.* 2001. Randomized study of antibodies to IFN-gamma and TNF-alpha in secondary progressive multiple sclerosis. *Mult. Scler.* **7:** 277–284.

66. Skurkovich, B. & S. Skurkovich. 2003. Anti-interferon-gamma antibodies in the treatment of au-

toimmune diseases. *Curr. Opin. Mol. Ther.* **5:** 52–57.

67. Hommes, D.W., T.L. Mikhajlova, S. Stoinov, *et al.* 2006. Fontolizumab, a humanised anti-interferon gamma antibody, demonstrates safety and clinical activity in patients with moderate to severe Crohn's disease. *Gut* **55:** 1131–1137.

68. Cannon G.W., S.H. Pincus, R.D. Emkey, *et al.* 1989. Double-blind trial of recombinant gamma-interferon versus placebo in the treatment of rheumatoid arthritis. *Arthritis Rheum.* **8:** 964–973.

69. Sigidin, Y.A., G.V. Loukina, B. Skurkovich & S. Skurkovich. 2001. Randomized, double-blind trial of anti-interferon-gamma antibodies in rheumatoid arthritis. *Scand. J. Rheumatol.* **30:** 203–207.

70. Robak, E., P. Smolewski, A. Wozniacka, *et al.* 2004. Relationship between peripheral blood dendritic cells and cytokines involved in the pathogenesis of systemic lupus erythematosus. *European Cytokine Network* **15:** 222–230.

71. Machold, K.P. & J.S. Smolen. 1990. Interferon-gamma induced exacerbation of systemic lupus erythematosus. *J. Rheumatol.* **17:** 831–832.

72. Theofilopoulos, A., S. Koundouris, D. Kono & B. Lawson. 2001. The role of IFN-gamma in systemic lupus erythematosus: a challenge to the Th1/Th2 paradigm in autoimmunity. *Arthritis Res.* **3:** 136–141.

# Interferon Lambda as a Potential New Therapeutic for Hepatitis C

**Dennis M. Miller,[a] Kevin M. Klucher,[a] Jeremy A. Freeman,[a] Diana F. Hausman,[b] David Fontana,[a] and Doug E. Williams[a]**

[a]ZymoGenetics, Inc., Seattle, Washington, USA, [b]Oncothyreon, Seattle, Washington, USA

Interferon lambdas (IFN-λ) are Type III interferons with biological activity, including induction of antiviral genes, similar to Type I IFNs, but signal through a distinct receptor complex. The expression pattern for the IFN-λ receptor is more cell specific than the widely distributed IFN-α receptor, suggesting *in vivo*, IFN-λ may have fewer side effects than IFN-α, such as less hematologic toxicities. A PEGylated form of IFN-λ (PEG-rIL-29) was well tolerated in animals and did not result in hematologic toxicity. Clinical data from initial studies of PEG-rIL-29 has demonstrated antiviral effects in patients with hepatitis C without producing hematologic toxicity. These preclinical and early clinical data support PEG-rIL-29 as a potential new therapeutic agent for treatment of patients with hepatitis C.

*Key words:* interferon-λ; hepatitis C; cytokines; PEG-IFN-λ

## Introduction

Interferon-lambda 1 (IFN-λ1), also known as interleukin-29 (IL-29), is a member of the recently described Type III IFN family.[1,2] There are three related cytokines that are members of the Type III IFN family, consisting of IL-28A, IL-28B, and IL-29 that are also known as IFN-λ 2, 3, and 1, respectively.[3-5] The Type III IFN have biological activities similar to Type I IFN, yet share very little sequence homology. Type III IFN signal through a different receptor complex than that used by the Type I IFN, further adding to the notion that Type III IFN are structurally distinct from the Type I IFN.

Type I IFN, such as IFN-alpha (IFN-α) and IFN-beta (IFN-β), regulate host responses to foreign pathogens by controlling both the innate and adaptive immune systems.[6,7] Similar to IFN-α, IFN-λ1 is induced in response to viral infection and has broad antiviral activity in preclinical studies, including inhibition of viral RNA replication in a hepatitis C virus (HCV) replicon model.[8-13] Because IFN-λ and the Type I IFN induce the same gene subsets, they are expected to have similar therapeutic effects on a cellular level, including on antiviral, antiproliferative, and apoptotic activities (Fig. 1). However, since IFN-λ binds to and signals through a unique cell-surface heterodimeric receptor, differences in the receptor expression patterns between these IFN families could lead to different effects *in vivo*, especially with regards to side effects from therapeutic uses.[13,14] The receptor for Type III IFN is composed of two chains, both of which are required for a functional receptor complex: IL-28Rα, which is unique to this receptor complex, and IL-10Rβ, which is shared among several Class II cytokine receptors.[1,2,4] In this paper, the heterodimeric receptor for the Type III IFN will be referred to as the IFN-λ receptor.

Expression of the IFN-λ receptor is more cell specific than that of the Type I IFN receptor, with expression of the IFN-λ receptor observed primarily on epithelial cells and being notably absent on most hematopoietic cells.[13,14] These observations suggest that

Address for correspondence: Dennis M. Miller, Ph.D., ZymoGenetics, Inc., 1201 Eastlake Avenue, East Seattle, WA 98102. Voice: 206-428-2754; fax: 206-442-6699. millerd@zgi.com

Cytokine Therapies: Ann. N.Y. Acad. Sci. 1182: 80–87 (2009).
doi: 10.1111/j.1749-6632.2009.05241.x © 2009 New York Academy of Sciences.

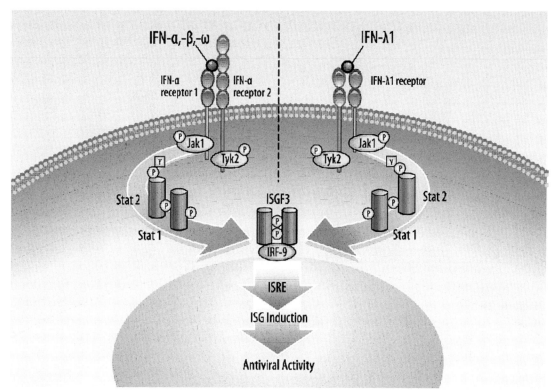

**Figure 1.** Signaling pathways of the Type I and Type III interferons. Figure reprinted with permission of ZymoGenetics, Inc.

systemic administration of Type III IFN may be associated with fewer adverse effects than from Type I IFN, such as less hematological toxicity.

ZymoGenetics, Inc. is evaluating a PEGylated Type III IFN, PEG-rIL-29, as a potential alternative to PEG-IFN-α for the treatment of chronic HCV infection. The current standard of care for HCV infection includes up to 48 weeks of treatment with PEG-IFN-α and ribavirin. This treatment can be associated with significant toxicity, leading to dose reductions, poor compliance with treatment, and avoidance of therapy by patients. A therapeutic compound with similar efficacy to PEG-IFN-α and an improved tolerability and safety profile is expected to increase patient adherence to prescribed therapy and ultimately lead to improved patient outcomes. Preclinical pharmacokinetic and toxicology studies, *in vivo* animal studies, and early clinical data suggest that PEG-rIL-29 may provide an alternative treatment option for patients with chronic hepatitis C.

## Preclinical Studies

### Effects *In Vitro*

Several published reports have described the similarity in gene expression patterns observed with treating responsive cells with Type I and Type III IFN *in vitro* antiviral activity of PEG-rIL-29 has been shown in models of HCV replication, including sub- and full-genomic replicon models, as well as other model systems sensitive to antiviral effects of IFN, such as encephalomyocarditis virus (EMCV).[1,3,5,10–13] The *in vitro* antiviral activity of PEG-rIL-29 is generally similar to that of PEG-IFN-α.[13]

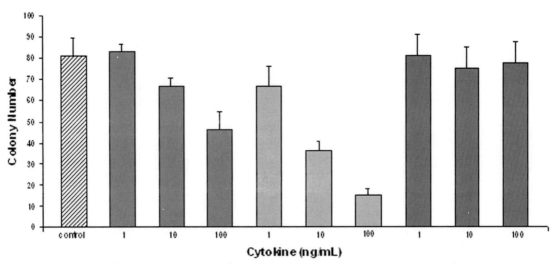

**Figure 2.** Lack of inhibition of colony forming unit (CFU) and granulocyte-macrophage (GM) colony formation by PEG-IFNλ: Comparison to PEG-IFN-α. Data are mean and standard deviation (n = 3 per group). Group 1 = PEG-IFN-α, Group 2 = PEG-IFNα-2b, Group 3 = PEG-rIL-29, Control = no cytokine.

However, comparisons of the *in vitro* potency estimates varied considerably among the model systems and endpoints used, ranging from relative potency differences of 30-fold less to 10-fold greater potency for PEG-rIL-29 compared to PEG-IFN-α. One possible reason for the relative potency differences is the varying expression levels of the two independent receptor complexes used by either IFN. Because these models reveal only a portion of the presumed *in vivo* antiviral effects of IFN, it was difficult to use these data to predict the dose levels of PEG-rIL-29 that would be associated with *in vivo* antiviral activity in patients.

The IFN-α receptor is expressed on all peripheral blood leukocytes (PBLs), including T and B cells, natural killer (NK) cells, and myeloid-derived cells, including monocytes. In contrast, the receptor for IL-29 is not highly expressed on NK or T cells, but receptor expression (as measured by mRNA) is detected on B cells.[13] Furthermore, the receptor for IL-29 was not detected on hematopoietic progenitor cells, suggesting the myelosuppressive effects of IFN-α may not be evident with PEG-rIL-

29. Figure 2 shows a dose-dependent inhibition of granulocyte-macrophage (GM) colony formation by two forms of PEG-IFN-α which contrasts to the lack of effects observed using PEG-rIL-29. Similar differences between PEG-rIL-29 and Type I IFN are observed for the red cell lineage (colony-forming unit-erythroid [CFU-E] formation, data not shown). These data suggest that PEG-rIL-29 will not produce the reported cytopenias (e.g., neutropenia) associated with PEG-IFN-α treatment *in vivo* and provide support to the concept that differences in receptor expression for PEG-rIL-29 may lead overall to fewer side effects.

## Effects *In Vivo*

The nonclinical safety profile of the PEG-IFN-α was characterized primarily in monkeys, with the main toxicology signals observed being a "sickness behavior" that may be analogous to the flu-like symptoms experienced by patients receiving PEG-IFN-α as well as changes in hematology parameters, possibly related

**TABLE 1.** Comparison of Preclinical Toxicology: PEG Interferon-alpha (IFN-α) versus PEG-IFN-λ

|  | PEG-IFN-α | PEG-IFN-λ |
|---|---|---|
| Dose (+ 50 mg/kg Ribavirin) | 0.45 mg/kg 2/wk | 2.5 mg/kg 2/wk |
| Mortality | 17% | None |
| Clinical observations | "Sickness behavior" | Normal |
| Body Weight | Decreased | No effect |
| RBC and WBC levels | Decreased | No effect |
| Serum Protein levels | Decreased total protein, albumin and globulin | Normal levels total protein, albumin, globulin |
| Bone Marrow | Hypocellularity | No effect |
| Thymus | Atrophy | No effect |

to the observed bone marrow hypocellularity (Table 1).[15] Safety studies of PEG-rIL-29 in monkeys were remarkable for the relatively few toxicities observed, including no evidence of a "sickness behavior" or of changes in hematology, including no apparent pathological effects on the bone marrow.

Consistent with the different effects on white cells counts observed *in vivo* between the PEG-IFN-α and PEG-rIL-29, *in vivo* pharmacology studies conducted in cynomolgus monkeys showed that PEG-rIL-29 did not induce expression of the antiviral genes Protein kinase R (*PkR*), Myxovirus resistance (*Mx*), and 2′5′oligoadenylate synthetase (*OAS*) in circulating peripheral blood leukocytes. In contrast, all three markers were induced in PBLs following IFN-α-2b (Intron®-A) dosing in monkeys.

Increased levels of antiviral genes, such as *OAS*, were observed in serial liver biopsies collected from monkeys treated with PEG-rIL-29, consistent with the presence of the receptor for PEG-rIL-29 on hepatocytes. The level of antiviral gene expression in the liver was generally comparable to that observed from monkeys treated with IFN-α.

*In vivo* modeling to evaluate antiviral efficacy is challenged by the lack of an appropriate model for hepatitis C. As a result, biomarkers of activity were used to estimate *in vivo* effects. $\beta_2$-microglobulin (B2M) is an extracellular component of the major histocompatability complex (MHC) Class I molecule that presents intracellular antigens to CD8$^+$ T cells. Expression of MHC Class I genes are induced by Type I IFN, and serum B2M increases following IFN dosing.[16–18] Changes in serum B2M were evaluated in repeat-dose pharmacodynamic, pharmacokinetic, and toxicology studies of PEG-rIL-29 in monkeys. Serum B2M was elevated following a single PEG-rIL-29 dose. Repeated dosing maintained the elevated level of serum B2M, which returned to baseline after the cessation of dosing. Across the studies conducted in cynomolgus monkeys, doses of at least 0.03 mg/kg induced increases in serum B2M. No change in serum B2M was observed at the lowest PEG-rIL-29 dose tested, 0.015 mg/kg. This dose was therefore considered the no observed effect level (NOEL) in cynomolgus monkeys for pharmacological changes and helped to form the basis of initial dose selection for clinical testing.

Finally, pharmacokinetic (PK) data were obtained from monkeys and used to estimate PK parameters in humans. The PK data in monkeys was observed to be linear across a wide dose range and estimates of the terminal half-life in monkeys were consistent with reported values for the PEG-IFN-α in monkeys, suggesting the weekly dosing in hepatitis C patients was feasible.

In summary, preclinical *in vivo* studies have demonstrated that PEG-rIL-29 induces cellular effects similar to the Type I IFN. The receptor distribution profile for Type III IFN is distinct from that of Type I, which results in cell-type-specific response pattern. Epithelial cell types are responsive to either IFN-λ or IFN-α; however, hematopoietic cells show little response to IFN-λ *in vitro* or *in vivo*. Administration of PEG-rIL-29 was well tolerated in animals, with no significant IFN-related clinical toxicities and no suggestion of additional anemic effects in combination with ribavirin. PK results were consistent with a PEGylated cytokine and suggest that weekly dosing is achievable. *in vivo* potency estimates were highly variable: the estimated NOEL for B2M in monkey was 15 μg/kg, which equates to a human equivalent dose of 5 μg/kg.

## Early Clinical Data

PEG-rIL-29 has been investigated in two clinical trials. A Phase 1a study evaluating single doses of PEG-rIL-29 was conducted in healthy subjects; the results of this study were used to select a pharmacologically active dose levels for further evaluation. The second Phase 1b study evaluated PEG-rIL-29 repeated doses administered every 2 weeks or weekly for 4 weeks as a single agent or in combination with ribavirin in patients with chronic HCV infection. Interim results from this study indicate that PEG-rIL-29 has antiviral activity and a favorable safety profile when administered as a single agent and in combination with ribavirin.

## Phase 1a Study in Healthy Subjects

ZymoGenetics, Inc. conducted a Phase 1, randomized, blinded, placebo-controlled dose-escalation study of single doses of PEG-rIL-29 administered to healthy subjects by SC injection. The objectives of this study were to evaluate the safety, tolerability, pharmacokinetics, and biologic activity of a single dose of PEG-rIL-29. Subjects were randomized in a 5:1 ratio to receive PEG-rIL-29 or placebo, administered subcutaneously on Day 1, and were continuously evaluated for safety for the first 48 h after dosing, as well as on Days 4, 8, 15, 29, and 59. Study assessments included documentation

of adverse events and laboratory measures, including cardiac measures. Samples to detect the presence of anti-PEG-rIL-29 antibodies were collected through Day 59. In addition to safety data, the pharmacokinetics of PEG-rIL-29 and serum levels of B2M were assessed.

A total of 20 subjects were enrolled in this study; 17 subjects were treated with PEG-rIL-29 (5 each at the 0.5, 1.5, and 5 μg/kg dose levels, and 2 at the 7.5 μg/kg dose level) and 3 subjects were treated with placebo. PEG-rIL-29 was well tolerated up to 5 μg/kg.

PEG-rIL-29 administration was not associated with fever, fatigue, or significant hematologic or cardiac effects and, with the exception of 1 subject treated at 7.5 μg/kg, was not associated with flu-like symptoms.[19] The primary safety observations consisted laboratory based measures, including reversible dose-related increases in liver transaminases in some subjects, which occurred without associated increases in bilirubin, and mild decreases in serum fibrinogen (mild decreases in fibrinogen were also observed in cynomolgus monkeys). There were no significant changes in any of the cardiac parameters evaluated. The dose limiting toxicity in this study was an increase in liver transaminases, with Grade 3 alanine aminotransferase (ALT) occurring in 1 of 2 subjects treated at 7.5 μg/kg.

PEG-rIL-29 demonstrated dose-dependent biologic activity with induction of increases in serum B2M starting at the 1.5 μg/kg dose. The increased serum B2M occurred at a dose level below the predicted pharmacologic dose range (>5 μg/kg) from the monkey, suggesting that PEG-rIL-29 is more potent in humans than predicted.

The pharmacokinetics of single doses of PEG-rIL-29 were consistent with weekly dosing. $T_{max}$ for individual subjects ranged from 4 to 72 h across all dose levels, and the estimated range for the terminal half-life was 50 to 80 h.

Results of this study demonstrated that single doses of PEG-rIL-29 was not associated with fever or fatigue, irritability or insomnia, injection site reactions, hematologic changes,

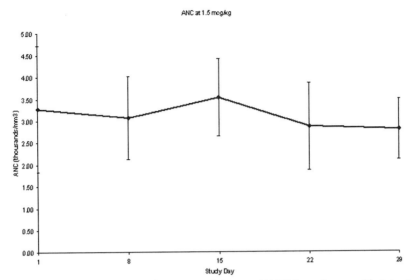

**Figure 3.** No decrease in absolute neutrophil count (ANC) from four weekly injections of PEG-rIL-29 in patients with hepatitis C. Data are mean and standard deviation ($n = 6$/time point).

or antibody development. Reversible dose-dependent effects included mild changes in coagulation parameters (Grade 1 decreases in fibrinogen) and asymptomatic increases in ALT/AST (aspartate aminotransferase) in some of the subjects treated at 5 and 7.5 μg/kg, which were not accompanied by increased bilirubin.

## Phase 1b Study in Subjects with Relapsed Chronic Hepatitis C

Based on the results of the Phase 1a study in healthy subjects, a Phase 1b study is ongoing to evaluate the safety and tolerability and antiviral activity of up to 4 weeks of treatment with PEG-rIL-29 as a single agent and in combination with ribavirin. Subjects arc cligible for this study if they have chronic genotype 1 HCV and have relapsed following treatment with PEG-IFN-α in combination with ribavirin. Evaluation in this particular population is appropriate for a new IFN molecule as the treatment is more likely to demonstrate antiviral activity in this population. A starting dose of 1.5 μg/kg was

chosen based on observed increases in B2M in the Phase 1a study.

Study results to date support the Phase 1a finding of pharmacologic activity being observed at 1.5 μg/kg dose level.[20] PEG-rIL-29, administered at 1.5 μg/kg per week resulted in 6 of 6 patients experiencing > 2-Log drop in HCV RNA and 4 of 6 patients with absolute HCV RNA levels of < 1000 IU/mL at Day 29 (Fig. 2). PEG-rIL-29 was well tolerated over a 4-week treatment period, with no evidence of hematological toxicity, including no neutropenia (absolute neutrophil count (ANC) $> 1.5 \times 10^3/\mu L$ for all patients: Fig. 3), thrombocytopenia, or anemia, and no treatment-related fever. Reports of flu-like symptoms have been minimal; reversible elevations in transaminases have been noted in several subjects. Enrollment in this study is ongoing.

## Conclusions

Our studies show that the IFN-λ receptor has cell-specific distribution, similar biological effects to Type I IFN on receptor expressing cells, *in vitro* antiviral activity similar to the Type I IFN, and little direct effects on hematopoietic cells. *in vivo* animal studies and early clinical studies with PEG-rIL-29 have not revealed a high degree of the classic IFN-α related side effects. In particular, hematopoietic toxicity, such as neutropenia or anemia, has not been observed following short-term administration of PEG-rIL-29. Reversible increases in liver transaminases have been observed in these early clinical studies of PEG-rIL-29 and have been the principal safety finding to date. Importantly, these early studies provide clear evidence of biological activity, most notably in the form of increases in serum B2M and decreases in serum HCV RNA.

*In vitro* systems were useful to characterize the qualitative nature of response in terms of IFN-stimulated gene (*ISG*) induction, receptor distribution, and cell-specific activity. While these systems allowed for observation of some of the direct antiviral effects, indirect effects such as may occur through enhanced the impact of enhanced MHC expression were harder to capture. Such systems were of limited usefulness for *in vivo* potency predictions, as was borne out by the observation of clinical activity occurred at lower doses than predicted from the *in vitro* human and *in vivo* animal data.

These data suggest that further clinical testing of PEG-rIL-29 is warranted in a larger Phase 2 setting. Especially valuable will be direct comparisons of PEG-rIL-29 to a Type I IFN in longer term studies to allow a more robust evaluation of the *in vivo* safety and activity profiles of these two structurally distinct IFN families.

## Acknowledgments and Disclosures

The authors wish to acknowledge the contributions of Karen Wisont, Grace Sweet, and Molly Bernard in preparing this manuscript for publication.

All authors are current or former employees of ZymoGenetics, Inc.

## Conflicts of interest

The authors declare no conflicts of interest.

# References

1. Sheppard, P. *et al.* 2003. IL-28, IL-29 and their class II cytokine receptor IL-28R. *Nat. Immunol.* **4:** 63–68.
2. Kotenko, S.V. *et al.* 2003. IFN-lambdas mediate antiviral protection through a distinct class II cytokine receptor complex. *Nat. Immunol.* **4:** 69–77.
3. Ank, N., H. West & S.R. Paludan. 2006. IFN-lambda: novel antiviral cytokines. *J. Interferon Cytokine Res.* **26:** 373–379.
4. Kotenko, S.V. & R.P. Donnelly. 2006. The IFN-lambda family (IL-28/29). *Anti-Inflammatory & Anti-Allergy Agents in Medicinal Chemistry* **5:** 279–285.
5. Li, M. *et al.* 2009. Interferon-lambdas: the modulators of antivirus, antitumor, and immune responses. *J. Leukoc. Biol.* **86:** 23–32.
6. Samuel, C.E. 2001. Antiviral actions of interferons. *Clin. Microbiol. Rev.* **14:** 778–809.
7. Stark, G.R. *et al.* 1998. How cells respond to interferons. *Annu. Rev. Biochem.* **67:** 227–264.
8. Mihm, S. *et al.* 2004. Interferon type I gene expression in chronic hepatitis C. *Lab. Invest.* **84:** 1148–1159.
9. Meager, A. *et al.* 2005. Biological activity of interleukins-28 and -29: comparison with type I interferons. *Cytokine* **31:** 109–118.
10. Robek, M.D., B.S. Boyd & F.V. Chisari. 2005. Lambda interferon inhibits hepatitis B and C virus replication. *J. Virol.* **79:** 3851–3854.
11. Zhu, H. *et al.* 2005. Novel type I interferon IL-28A suppresses hepatitis C viral RNA replication. *Virol. J.* **2:** 80.
12. Ank, N. *et al.* 2006. Lambda interferon (IFN-lambda), a type III IFN, is induced by viruses and IFNs and displays potent antiviral activity against select virus infections in vivo. *J. Virol.* **80:** 4501–4509.
13. Doyle, S.E. *et al.* 2006. Interleukin-29 uses a type 1 interferon-like program to promote antiviral responses in human hepatocytes. *Hepatology* **44:** 896–906.
14. Sommereyns, C. *et al.* 2008. IFN-lambda (IFN-lambda) is expressed in a tissue-dependent fashion and primarily acts on epithelial cells in vivo. *PLoS Pathog.* **4:** e1000017.
15. *Summary Basis of Approval for Peg-Intron (Peginterferon alfa 2b), Toxicologist's Review.* FDA website, 2001. pp. 1–60.
16. Buraglio, M. *et al.* 1999. Recombinant human interferon-beta-1a (Rebif®) vs recombinant

interferon-beta-1b (Betaseron®) in healthy volunteers. *Clin. Drug Invest.* **18:** 27–34.

17. Carreno, V. *et al.* 2000. A phase I/II study of recombinant human interleukin-12 in patients with chronic hepatitis B. *J. Hepatol.* **32:** 317–324.

18. Witt, P.L. *et al.* 1993. Pharmacodynamics of biological response in vivo after single and multiple doses of interferon-beta. *J. Immunother.* **13:** 191–200.

19. Hausman, D.F. *et al.* 2007. Abstract 67: A phase 1, randomized, blinded, placebo-controlled, single-dose, dose-escalation study of PEG-interferon lamda (PEG-rIL-29) in healthy subjects HEP DART 2007. *Frontiers in Drug Development for Viral Hepatitis* **3**(Supp 2): 71–72.

20. Lawitz, E. *et al.* 2008. Phase 1b dose-escalation study of PEG-interferon-lambda (PEG-rIL-29) in relapsed chronic hepatitis C patients; abstract 170. *Hepatology* **48**(4 Suppl): 385A.

# Cytokines as Targets
# for Anti-inflammatory Agents

## Larry W. Moreland

*University of Pittsburgh, Pittsburgh, Pennsylvania, USA*

**Well over a decade ago a central role of tumor necrosis factor (TNF) was first described in patients with rheumatoid arthritis (RA) when remarkable clinical benefit was demonstrated in patients with refractory disease were treated with using either a monoclonal antibody or a soluble receptor fusion protein. There are now five anti-TNF agents approved by regulatory agencies for treating RA. Identifying which RA patients will have a meaningful clinical response (improvement in outcomes measures such as ACR 20, DAS score, remission, etc.) when used as monotherapy, or in combination with other immunosuppressive agents remains a major research effort. Also, attention has focused on the potential adverse events that can be seen with these therapies; an increase in opportunistic infections being the most clearly linked adverse event. These anti-TNF therapies have revolutionized the clinicians' ability to make a significant impact in RA, a disease that has significant excess morbidity and mortality.**

*Key words:* **rheumatoid arthritis; biologics; cytokines; anti-inflammatory agents**

## Overview of Disease

The past decade has seen the introduction of an unprecedented number of new therapeutic options to treat rheumatoid arthritis (RA), a disease associated with premature death and excess mortality.[1–74] An improved understanding of the pathophysiology of RA has led to several key changes in the approach to therapy. The treatment of RA has markedly changed over the recent years because of new data supporting the use of combination therapy[1–4,29]; the introduction of new disease modifying anti-rheumatic drugs (DMARDs) such as leflunomide,[7] etanercept,[8–11] infliximab,[12–15] adalimumab,[16,26] abatacept,[38] rituximab[28,30] and anakinra.[17,18] Among traditional DMARDs, methotrexate (MTX) has emerged as the preferred first-line agent of U.S. rheumatologists, displaying favorable toxicity and efficacy profiles.[19,20] Other oral DMARDs that are often used include hydroxychloroquine, sulfasalazine, and leflunomide. The use of DMARDs early, and in many patients, in combination, is highly effective.

Many of these drugs now offer the real possibility of disease remission in a large proportion of treated patients. This has resulted in the prospect of patients having their symptoms suppressed, their joints preserved from destruction, and their function improved. These prospects all come with the possibility of fewer side effects than were seen with older treatment regimens.

Since RA is a heterogeneous disease, it is evident that the specific aims of its management will vary among individual patients, depending on the aggressiveness of the disease, age and life status at onset, and current life status as it is affected by the signs and symptoms of inflammation, decreased physical function, work disability, destruction of specific joints, and social and emotional coping capacity.

## Anti-TNF Therapies

Recent advances in biotechnology have allowed for the commercial development of

Address for correspondence: Larry W. Moreland, M.D., University of Pittsburgh, 3500 Terrace Street, Biomedical Science Tower S711, Pittsburgh, PA 15261. Voice: 412-648-0148. lwm5@pitt.edu

Cytokine Therapies: Ann. N.Y. Acad. Sci. 1182: 88–96 (2009).
doi: 10.1111/j.1749-6632.2009.05072.x © 2009 New York Academy of Sciences.

agents that specifically target inflammatory cytokines. In the last 10 years, the FDA approved eight biological agents (etanercept, infliximab, anakinra, adalimumab, certolizumab, golimumab, abatacept, and rituximab) for the treatment of RA. With the availability of five TNF inhibitors (infliximab, etanercept, adalimumab, certolizumab, and golimumab) for the treatment of RA, the major task facing physicians is finding ways to predict response to these agents as well as that to other standard DMARDs and combinations. The unique mode of action of these anti-TNF agents and their relatively rapid onset of action makes them attractive agents for use as "induction" therapies in combination regimens with traditional DMARDs.

Since their introduction into the clinic in 1998, inhibitors of the proinflammatory cytokine TNF have had a dramatic impact on physicians' therapeutic approach to patients with RA and other autoimmune systemic inflammatory diseases. In patients with RA, the ability of TNF inhibitors to improve the signs and symptoms of disease considerably and to enhance functional status and attenuate the progression of joint damage has dramatically altered our treatment options and goals. In refractory disease, the addition of these anti-TNF therapies to MTX treatment provides additional benefits to patients with persistent active disease.

Controlled clinical trials have demonstrated that measurable radiographic progression of joint damage can be slowed or significantly reduced using currently available therapies. Improvement of signs, symptoms and function also occurred. In particular, combination of MTX with a TNF inhibitor appears to slow the progression of radiographic progression in most patients.[26,36,43] Recent data support treating patients with early RA more aggressively in order to slow or prevent structural damage,[21,53] maintain good function and avoid work disability, and reduce mortality.[34,41,54] Recent data from several placebo-controlled trials support treating patients with early RA with combination therapies such as MTX and anti-TNF therapies in order to slow or prevent structural damage.[26,36,43] However, many patients who are treated with just MTX as monotherapy also can have remission of their disease and slowing of radiographic progression. Thus, what is urgently needed are biomarkers that allow clinicians to select which RA patients need treatment with monotherapy (e.g., MTX) versus combination DMARDs (e.g., "triple therapy" vs. anti-TNF plus MTX).

Recent data from the BeSt and COMET trials provide very interesting data regarding the possibility of inducing long-term remission in a subset of RA patients who are treated early with a combination of anti-TNF agents plus MTX.[33,73,74] This illustrates the need to better measure the immune abnormalities in each individual RA patient so we can more effectively apply the appropriate treatments, that is, personalized medicine.

Although all first three TNF inhibitors approved (etanercept, infliximab, and certolizumab) and on the market the longest are clinically effective, there are differences among them. Infliximab and adalimumab are specific for TNF-α; etanercept binds both TNF-α and lymphotoxin-α. Although all bind with high affinity, the avidity and, hence, duration of binding may be greater for the MABs. Effector functions such as induction of cell lysis and apoptosis can be demonstrated by *in vitro* studies with the MABs but not the soluble receptor; the *in vivo* relevance of this is uncertain. Given intravenously, infliximab has a high peak concentration followed by steady-state elimination, whereas etanercept and adalimumab, because they are given subcutaneously, have more flat pharmacokinetic profiles. All agents are effective in controlling the signs and symptoms of RA. Whether these differences in characteristics between the agents will result in differences in other outcomes in RA patients (e.g., effect on radiographic changes), efficacy in other diseases, and toxicities, remains largely to be shown.

**TABLE 1.** Potential Toxicities of TNF Inhibitors

- Injection site reactions or infusion reactions
- Infections: bacterial and opportunisitic
- Demelinating disorders
- Lymphoproliferative disorders
- Congestive heart failure
- Cytopenias
- Hepatotoxicity
- Vasculitis
- Autoantibody production (ANA, DNA, etc.)
- Antibodies to administered proteins (immunogenicity)

There is limited data that RA patients not responding, or responding initially, and then losing the positive clinical benefit, to one anti-TNF agent, may have some improvement in disease activity when treated with a different anti-TNF agent. This preliminary data needs to be further explored in well-controlled larger studies where potential mechanistic studies can be applied trying to better understand this heterogenecity in RA patients and the five anti-TNF agents.

Adverse events (Table 1) related to the use of TNF-α inhibitors can be grouped into those that are agent related versus those that are target related. Injection site and infusion reactions, and immunogenicity, vary depending on the particular agent. Increased predisposition to infections, development of malignancy, induction of autoimmune disorders, and associations with demyelinating disorders, myelosuppression, and worse outcome with congestive heart failure might be considered target-related adverse events. Any clinically effective TNF-α inhibitor might be associated with such adverse events, although the relative risk amount different agents may vary depending on characteristics such as target specificity, affinity or avidity, effector functions, and pharmacokinetic or pharmacodynamic considerations.

The central role that TNF plays in normal immunity, as well as the awareness of the adverse effects related to the use of other immunomodulatory treatments lead to potential concerns of TNF inhibitors. Significant interest and attention have been directed toward the prevalence of infections, serious infections, and opportunistic infections with the use of TNF inhibitors.[68–72,75] Compared with the general population, patients who have RA tend to have more infections. Also relevant is the higher risk of infection in patients with RA who have the most active and severe disease; this is also the subset of patients most commonly treated with TNF inhibitors.

The most common sites of infection in RA patients, such as those involving the respiratory tract, are the sites most often seen in the overall population. In postmarketing experiences, there has been careful attention to monitoring for serious infections in RA patients treated with TNF inhibitors. In addition to common infections, various opportunistic infections (e.g., *Pneumocystic carinii; Legionella;* tuberculosis (TB); atypical mycobacteria infections; listeriosis; coccidioidomycosis; histoplasmosis; and aspergillosis) have been reported in patients receiving therapy with TNF inhibitors. In many cases, the development of these infections reflects the overall incidence within the local community.[68]

Data from numerous animal studies have shown that TNF-α plays a critical role in defense against TB. Although there were few cases of TB during clinical trials that involved TNF inhibitors, postmarketing surveillance revealed a substantial number of cases. Most cases appear to be reactivation of latent TB. Although there may be a greater risk of TB and other granulomatous infections in patient receiving TNF inhibitors that are monoclonal antibodies, cases of TB have been seen with all agents. Thus, vigilance for such infections is required with all TNF inhibitor therapy. Despite the greater use of TNF inhibitors in the United States, a greater number of cases of TB have occurred outside this country, particularly in countries in which there is a higher prevalence of TB in the general population. About half the presentations of TB in patients taking TNF inhibitors were extrapulmonary or disseminated; this rate contrasts with that in the general population, in which 80% or more of patients who have TB present with pneumonia. Although screening for latent TB has been very effective, vigilance is still required, because cutaneous anergy is not uncommon in the populations of patients treated with TNF inhibitors and new cases of TB my arise during therapy.

Another potential adverse effect that may be seen with immunodulatory therapies is malignancy.[70] To date, TNF inhibitor therapy in patients with RA has not been associated with a greater risk of solid organ tumors. A twofold or greater risk of non-Hodgkin's lymphoma has been observed in patients with RA who are receiving TNF inhibitors, compared with an age and sex-matched population. However, this rate may approximate the baseline risk of the exposed RA population. Several studies have shown that patients with RA are at higher risk for lymphoma, particularly patients who have the most severe and active disease. Because this is the population of patients with RA most likely to receive TNF inhibitor therapy, more studies with longer follow-up time are needed to fully define any potential association between lymphoma and these agents.

An interesting aspect of treatment of patients with TNF inhibitors has been the development of autoantibodies. Although positive antinuclear antibodies (ANAs) are not uncommon in patients with RA, their prevalence increases to 50% or more in the patients treated with TNF inhibitors. Autoantibodies against double-stranded DNA (anti-DNA) also develop in about 10% of patients treated with TNF inhibitors. Such autoantibodies are much less common in RA and are more specific for SLE than is the generic ANA. What appears to be drug-induced lupus (or "lupus-like illness") related to TNF inhibitors has developed in some patients, but this is uncommon and certainly far less frequent than the development of autoantibodies. Clinicians should be aware of this phenomenon of autoantibodies with TNF inhibitor therapy, but a positive ANA or anti-DNA test result in a patient with RA who is receiving a TNF inhibitor does not mean that the patient had SLE rather than RA or that the patient had

drug-induced lupus. In the absence of clinical symptoms thought to be directly related to the agent, it is not necessary to discontinue TNF inhibitor therapy solely because of the presence of these autoantibodies.

Exacerbation of previously quiescent multiple sclerosis and new-onset demyelinating neurologic disease have been reported. The number of patients affected is not known. Although a causal relationship has not been established, the fact that another TNF antagonist, lenercept, worsened symptoms in patients with multiple sclerosis renders the association plausible.

Immunogenicity of administered proteins can result in neutralization of the biologic agent.[59-65] Limited data is available on the prevalence of real clinical significance of the immunogenicity of TNF inhibitors in RA patients. Further details of the assays used are in the manufacturers package inserts. Further research is needed to determine what role, if any, antibodies to TNF inhibitors play in loss of efficacy in treated patients. Key is this area of research is developing standardized assays to measure TNF inhibition and the role of antibodies to TNF inhibitors in altering TNF inhibition. Are the measurable antibodies neutralizing or not? Are higher measurable quantities of the antibodies clinically significant? Does the route of administration (i.v vs. s.q.) of the anti-TNF agent, and the concomitant administration of other immunosuppressive agents (e.g., MTX) have an impact on the development of and clinical significance of these autoantibodies? Should we ultimately be measuring patients' serum for levels of these antibodies and serum drug concentrations?

Serum TNF-$\alpha$ levels are elevated in patients with heart failure and associated with decreased cardiac contractility. Initial reports on anti-TNF therapy for heart failure were encouraging. However, subsequent studies of etanercept and infliximab in heart failure were stopped early because of lack of evidence of benefit and, in the case of infliximab, increased mortality.

No specific laboratory monitoring is currently required by regulatory agencies during therapy with TNF-α inhibitors. Because of the rare occurrence of myelosuppression and concern about the risk of infections, clinicians may obtain intermittent assessment of the CBC. Careful monitoring of patients for any sign and symptom of infection, demyelinating disease, and malignancy is suggested during treatment with all TNF-α inhibitors.

## IL-1 Inhibition

Strategies for inhibition of IL-1 include the development of genetically-engineered recombinant forms of IL-1 RA and IL-1 receptors. Recombinant IL-1 RA (Anakinra) is the only IL-1 antagonist approved for treating RA.[17,18] It is administered by daily subcutaneous injections. Although no head-to-head comparative studies against anti-TNF therapies have been performed, its efficacy and use in RA is limited compared to the anti-TNF therapies.

## B-cell Inhibitors

Until recently the T cell was believed to be the main immune cell involved in the abnormal self-antigen recognition that characterizes autoimmune diseases. Over the last few years, considerable evidence has emerged that B cells may also play a central role in the pathogenesis of autoimmune diseases by presenting antigens to T cells, providing costimulatory signals, secreting cytokines, and also by producing autoantibodies. B cells' critical role in T cell activation has justified the trial of several therapeutic agents.

CD20 is a surface antigen that is expressed only on mature B cells. Rituximab is a genetically engineered chimeric monoclonal antibody to anti-CD20 IgG1, which has been approved for the treatment of B cell lymphoma since 1997. Rituximab causes selective and rapid transient depletion of the CD20+ B cell population. B cell depletion by rituximab may be mediated by various mechanisms, including antibody-dependent cellular cytotoxicity,

complement-mediated B cell lysis, and induction of apoptosis.

Rituximab therapy induces profound peripheral blood B cell depletion for 6 months or more, and the therapy has also associated with a decrease in rheumatoid factor (RF) levels that was maintained after 24 weeks. Infections have not been more common with rituximab than with methotrexate alone. Infusion-related events, including transient hypotension, cough, pruritus, and rash were the most common side-effects during the first rituximab infusion.

The mechanisms by which Rituximab improves disease activity in RA potentially can be decreasing the production of cytokines.[28,30] Several other agents that inhibit B cell function are now in clinical trials.

## Co-stimulatory Blockade

Autoreactive T cells stimulated by one or more unidentified autoantigens presumably play a central role in triggering RA. Activation of T cells requires both antigen presentation and costimulation by the antigen-presenting cells (APCs). The first signal is generated when an antigen/MHC class II complex binds to the specific T cell receptor. Although essential for initiating T cell activation, this signal is not sufficient to induce a productive immune response. A second, costimulatory signal occurs when cell-surface molecules on the APC (CD80 or CD86) interact with a heterodimeric cell-surface protein receptor (CD28) on the T cell surface. Cytotoxic T lymphocyte-associated antigen 4 (CTLA-4), which is an important negative regulator of CD80/CD86—CD28, is a molecule that is transiently expressed at the surface of the T cells. CTLA-4 is expressed by T cells early after activation, but not on resting T cells, and transduces an inhibitory signal. As CTLA-4 binds to CD80/cd86 on the APC with a greater avidity than CD28, it also competes with CD28 binding to CD80/CD86, thereby leading to incomplete T cell activation. These observations led to the development of abatacept, CTLA-4-immunoglobulin (CTLA-4Ig), a soluble receptor construct composed of the extracellular domain of CTLA-4 and an IgG1 Fc piece.[8,37]

The mechanisms by which abatacept may improve disease activity in RA are still being investigated. However, a decrease in pro-inflammatory cytokines may be responsible for this clinical benefit.

## IL-6 Inhibition

Like TNF-α, IL-6 is a pleiotropic cytokine with a variety of biologic effects. IL-6 shares structural homology with a number of other cytokines, including IL-11, leukemia inhibitory factor, cardiotrophin-1, oncostatin M, and ciliary neurotrophic factor, together forming the IL-6 superfamily. These cytokines also produce biologic effects through similar receptor interactions and signaling pathways.

Both cell-bound and soluble IL-6Rs contribute to cellular activation. Cells expressing gp130 but no IL-6R can be stimulated by IL-6 through sIL-6R, and the biological effects of IL-6 can be blocked by inhibiting IL-6, IL-6R, or gp130. However, blocking IL-6 or sIL-6R will inhibit the biological effect of IL-6 alone, but inhibition of gp130 will impede the function of all the cytokines in the IL-6 superfamily.

Several pivotal trials evaluating tocilizumab (Mab to IL-6R) have been performed in RA patients.[57] Significant clinical benefit has been demonstrated when the agent has been studied as monotherapy, or in combination with MTX. Therapy related adverse events reported to date include transient decrease in WBC, transient elevation of liver enzymes, increased lipids, and reports of bacterial infections. As of this writing this agent is being reviewed by the FDA as a therapy for RA.

## Signal Transduction Inhibitors

There are several intracellular signaling pathways that facilitate communications within cells in response to a variety of stimuli (e.g., cytokines). These complex pathways are the target of new drug development for a variety of immune-mediated disorders. For example, mitogen-activated protein (MAP) kinase, janus-kinase (JAK), and spleen tyrosine (SYK) are three such targets that are actively being evaluated in clinical trials involving RA patients. To date, there has been disappointment with the development of MAP kinase inhibitors for RA.[54,55] However, pivotal large studies are now being reported with the JAK and SYK inhibitors.[56] Inhibitors of p38 kinase have largely failed in clinical trials, due to both lack of efficacy and adverse events. The degree of adverse events may reflect off-target effects or, conversely, may be a mechanism-related event subsequent to successful inhibition of p38.

## Summary

Remarkable progress has been made in the past decade with newer, targeted therapies for treating RA. These treatments have changed the course and face of RA outcomes. Further research is urgently needed to better understand the cause and pathogenesis of different RA subsets that will lead to individualized therapy. Now with eight, possibly nine in next few months, biological therapies approved to treat RA, there are numerous opportunities to perform cost effectiveness as well as cost comparative research with these new targeted biological therapies and with the traditional oral DMARDs.

### Conflicts of Interest

The author has been an investigator for several pharmaceutically sponsored studies over the past 15 years. Also, he has been a consultant for many biotech and pharmaceutical companies in the past.

## References

1. Boers, M., A.C. Verhoeven, H.M. Markusse, *et al*. 1997. Randomized comparison of combined

step-down prednisolone, methotrexate and sulfasalazine with sulfasalazine alone in early rheumatoid arthritis. *Lancet* **350:** 309–318.

2. Mottonen, T., P. Hannonen, M. Leirisalo-Repo, *et al.* 1999. Comparison of combination therapy with single-drug therapy in early rheumatoid arthritis: a randomized trial. *Lancet* **353:** 1568–1573.

3. O'Dell, J.R., C.E. Haire, N. Erikson, *et al.* 1996. Treatment of rheumatoid arthritis with methotrexate alone, sulfasalazine and hydroxychloroquine, or a combination of all three medications. *N. Engl. J. Med.* **334:** 1287–1291.

4. Landewe, R.B.M., M. Boers, A.C. Verhoeven, *et al.* 2002. COBRA combination therapy in patients with early rheumatoid arthritis. *Arthritis Rheum.* **46:** 347–356.

5. Mottonen, T., P. Hannonen, M. Korpela, *et al.* ; FIN-RACo Trial Group. 2002. Delay to institution of therapy and induction of remission using single-drug or combination-disease modifying antirheumatic drug therapy in early rheumatoid arthritis. *Arthritis Rheum.* **46:** 894–898.

6. Smolen, J.S., J.R. Kalden & D.L. Scott, *et al.* 1999. Efficacy and safety of leflunomide compared with placebo and sulphasalazine in active rheumatoid arthritis: a double-blind, randomized, multicentre trial. European Leflunomide Study Group *Lancet* **353:** 259–266.

7. Cohen, S., M. Schiff, A. Weaver, *et al.* 1999. Treatment of active rheumatoid arthritis with leflunomide compared to placebo and methotrexate. *Arch. Intern. Med.* **159:** 2542–2550.

8. Bathon, J.M., R.W. Martin, R.M. Fleischmann, *et al.* 2000. A comparison of etanercept and methotrexate in patients with early rheumatoid arthritis. *N. Engl. J. Med.* **343:** 1586–1593.

9. Moreland, L.W., M.H. Schiff, S.W. Baumgartner, *et al.* 1999. Etanercept therapy in rheumatoid arthritis. A randomized, controlled trial. *Ann. Intern. Med.* **130:** 478–486.

10. Weinblatt, M.E., J.M. Kremer, A.D. Bankhurst, *et al.* 1999. A trial of etanercept, a recombinant tumor necrosis factor receptor: Fc fusion protein, in patients with rheumatoid arthritis receiving methotrexate. *N. Engl. J. Med.* **340:** 253–259.

11. Moreland, L.W., S.W. Baumgartner, M.H. Schiff, *et al.* 1997. Treatment of rheumatoid arthritis with a recombinant human tumor necrosis factor receptor (p75)-Fc fusion protein. *N. Engl. J. Med.* **337:** 141–147.

12. Elliott, M.J., R.N. Maini, M. Feldmann, *et al.* 1994. Randomized double-blind comparison of chimeric monoclonal antibody to tumor necrosis factor-alpha (cA2) versus placebo in rheumatoid arthritis. *Lancet* **344:** 1105–1110.

13. Elliott, M.J., R.N. Maini, M. Feldmann, *et al.* 1993. Treatment of rheumatoid arthritis with chimeric monoclonal antibodies to tumor necrosis factor-alpha. *Arthritis Rheum.* **36:** 1681–1690.

14. Maini, R., E.W. St Clair, F. Breedveld, *et al.* 1999. Infliximab (chimeric anti-tumour necrosis factor-α monoclonal antibody) versus placebo in rheumatoid arthritis patients receiving concomitant methotrexate: a randomized phase III trial. *Lancet* **354:** 1932–1939.

15. Lipsky, P.E., D.M. Van Der Heijde, E.W. St Clair, *et al.* 2000. Infliximab and methotrexate in the treatment of rheumatoid arthritis. *N. Engl. J. Med.* **343:** 1594–1602.

16. Weinblatt, M., E. Keystone, D. Furst, *et al.* 2003. Adalimumab, a fully human anti-TNF-α monoclonal antibody, for the treatment of rheumatoid arthritis patients taking concomitant methotrexate: the ARMADA Trial. *Arthritis Rheum.* **48:** 35–45.

17. Bresnihan, B., J.M. Alvaro-Gracia, M. Cobby, *et al.* 1998. Treatment of rheumatoid arthritis with recombinant human interleukin-1 receptor antagonist. *Arthritis Rheum.* **41:** 2196–2204.

18. Jiang, Y., H.K. Genant, I. Watt, *et al.* 2000. A multicenter, double-blind, dose-ranging, randomized, placebo-controlled study of recombinant human interleukin-1 receptor antagonist in patients with rheumatoid arthritis. Radiologic progression and correlation of Genant and Larsen scores. *Arthritis Rheum.* **43:** 1001–1009.

19. Choi, H.K., M.A. Hernan, J.D. Seeger, *et al.* 2002. Methotrexate and mortality in patients with rheumatoid arthritis: a prospective study. *Lancet* **359:** 1173–1177.

20. Lee, D.M. & M.E. Weinblatt. 2001. Rheumatoid arthritis. *Lancet* **358:** 903–911.

21. Van Der Heide, A., J.W.G. Jacobs & J.W.J. Bijlsma. 1996. The effectiveness of early treatment with "second-line" antirheumatic drugs: a randomized, controlled trial. *Ann. Intern. Med.* **124:** 699–707.

22. Lard, L.R., M. Boers, A. Verhoeven, *et al.* 2002. Early and aggressive treatment of rheumatoid arthritis patients affects the association of HLA Class II antigens with progression of joint damage. *Arthritis Rheum.* **46:** 899–905.

23. Boers, M. 2001. Rheumatoid arthritis. Treatment of early disease. *Rheum. Dis. Clin. N. Am.* **27:** 405–414.

24. Moreland, L.W. & S.L. Bridges Jr. 2001. Early rheumatoid arthritis: a medical emergency? *Am. J. Med.* **111:** 498–500.

25. Boers, M. 2003. Add-on or step-up trials for new drug development in rheumatoid arthritis: A new standard? *Arthritis Rheum.* **48:** 1481–1483.

26. Breedveld, F.C., M.H. Weisman, A.F. Kavanaugh, *et al.* 2006. The PREMIER Study: A multicenter,

randomized, double-blind clinical trial of combination therapy with adalimumab plus methotrexate versus methotrexate alone or adalimumab alone in patients with early, aggressive rheumatoid arthritis who had not had previous methotrexate treatment. *Arthritis Rheum.* **54:** 26–37.

27. de Vries-Bouwstra, J.K., B. Dijkmans & F. Breedveld. 2005. Biologics in early rheumatoid arthritis. *Rheum. Dis. Clin. N. Am.* **31:** 745–762.

28. Edwards, J., L. Szczepanski, J. Szechinski, *et al.* 2004. Efficacy of B-cell-targeted therapy with rituximab in patients with rheumatoid arthritis. *N. Engl. J. Med.* **350:** 2572–2581.

29. Emery, P. 2006. Clinical review: Treatment of rheumatoid arthritis. *BMJ* **332:** 152–155.

30. Emery, P., R. Fleischmann, A. Filipowicz-Sosnowska, *et al.* 2006. The efficacy and safety of rituximab in patients with active rheumatoid arthritis despite methotrexate treatment. *Arthritis Rheum.* **54:** 1390–1400.

31. Emery, P. 2005. Adalimumab therapy: clinical findings and implications for integration into clinical guidelines for rheumatoid arthritis. *Drugs Today* **41:** 155–163.

32. Genovese, M.C., J. Becker, M. Schiff, *et al.* 2005. Abatacept for rheumatoid arthritis refractory to tumor necrosis factor α inhibition. *N. Engl. J. Med.* **353:** 1114–1123.

33. Goekoop-Ruiterman, Y.P.M., J.K. de Vries-Bouwstra, C.F. Allaart, *et al.* 2005. Clinical and radiographic outcomes of four different treatment strategies in patients with early rheumatoid arthritis (the BeST Study). *Arthritis Rheum.* **52:** 3381–3390.

34. Hyrich, K.L., D. Symmons, K.D. Watson & A.J. Silman. 2006. Comparison of the response to infliximab or etanercept monotherapy with the response to cotherapy with methotrexate or another disease-modifying antirheumatic drug in patients with rheumatoid arthritis. *Arthritis Rheum.* **54:** 1786–1794.

35. Keystone, E.C., A.F. Kavanaugh, J.T. Sharp, *et al.* 2004. Radiographic, clinical, and functional outcomes of treatment with adalimumab (a human anti-tumor necrosis factor monoclonal antibody) in patients with active rheumatoid arthritis receiving concomitant methotrexate therapy: a randomized, placebo-controlled, 52-week trial. *Arthritis Rheum.* **50:** 1400–1411.

36. Klareskog, L., D. Van Der Heijde, J.P. de Jager, *et al.* 2004. Therapeutic effect of the combination of etanercept and methotrexate compared with each treatment alone in patients with rheumatoid arthritis: double-blind randomised controlled trial. *Lancet* **363:** 675–681.

37. Kremer, J.M., R. Westhovens, M. Leon, *et al.* 2003. Treatment of rheumatoid arthritis by selective inhibition of T-cell activation with fusion protein CTLA4Ig. *New Engl. J. Med.*. **349:** 1907–1915.

38. Moreland, L.W., R. Alten, F. Van de Bosch, *et al.* 2002. Costimulatory blockade in patients with rheumatoid arthritis: a pilot, dose-finding, double blind, placebo-controlled clinical trial evaluating CTLA- 4Ig and LEA29Y eighty-five days after the first infusion. *Arthritis Rheum.* **46:** 1470–1479.

39. Quinn, M.A. & P. Emery. 2005. Potential for altering rheumatoid arthritis outcome. *Rheum. Dis. Clin. N. Am.* **31:** 763–772.

40. Quinn, M.A., P.G. Gonaghan, P.J. O'Connor, *et al.* 2005. Very early treatment with infliximab in addition to methotrexate in early, poor-prognosis rheumatoid arthritis reduces magnetic resonance imaging evidence of synovitis and damage, with sustained benefit after infliximab withdrawal: Results from a twelve-month randomized, double-blind, placebo-controlled trial. *Arthritis Rheum.* **52:** 27–35.

41. Smolen, J.S., C. Han, M. Bala, *et al.* 2005. Evidence of radiographic benefit of treatment with infliximab plus methotrexate in rheumatoid arthritis patients who had no clinical improvement. *Arthritis Rheum.* **52:** 1020–1030.

42. Smolen, J.S., D. Van Der Heijde, E.W. St Clair, *et al.* 2006. Predictors of joint damage in patients with early rheumatoid arthritis treated with high-dose methotrexate with or without concomitant infliximab. *Arthritis Rheum.* **54:** 702–710.

43. St Clair, E.W., D. Van Der Heijde, J.S. Smolen, *et al.* 2004. Combination of infliximab and methotrexate therapy for early rheumatoid arthritis. *Arthritis Rheum.* **50:** 3432–3443.

44. Van Der Heijde, D., L. Klareskog, V. Rodriguez-Valverde, *et al.* 2006. Comparison of etanercept and methotrexate, alone and combined, in the treatment of rheumatoid arthritis: two-year clinical and radiographic results from TEMPO Study, a double blind, randomized trial. *Arthritis Rheum.* **54:** 1063–1074.

45. Weyand, C.M. & J. Goronzy. 2006. T-cell-targeted therapies in rheumatoid arthritis. *Nat. Clin. Pract. Rheumatol.* **2:** 201–210.

46. Lard, L.R., H. Visser, I. Speyer, *et al.* 2001. Early versus delayed treatment in patients with recent-onset rheumatoid arthritis: comparison of two cohorts who received different treatment strategies. *Am. J. Med.* **111:** 446–451.

47. Yelin, E.H. & P.P. Katz. 2002. Focusing interventions for disability among patients with rheumatoid arthritis. *Arthritis Rheum.* **47:** 231–233.

48. Escalante, A. & I. del Rincon. 2002. The disablement process in rheumatoid arthritis. *Arthritis Rheum.* **47:** 333–342.

49. van Doornum, S., G. McColl & I.P. Wicks. 2002. Accelerated atherosclerosis. An extraarticular feature of rheumatoid arthritis? *Arthritis Rheum.* **46:** 862–873.

50. Wong, J.B., D.R. Ramey & G. Singh. 2001. Long-term morbidity, mortality and economics of rheumatoid arthritis. *Arthritis Rheum.* **44:** 2746–2749.

51. Bacon, P.A. & J.N. Townend. 2001. Nails in the coffin: Increasing evidence for the role of rheumatic disease in the cardiovascular mortality of rheumatoid arthritis. *Arthritis Rheum.* **44:** 2707–2710.

52. del Rincon, I., K. Williams, M.P. Stern, *et al.* 2001. High incidence of cardiovascular events in a rheumatoid arthritis cohort not explained by traditional cardiac risk factors. *Arthritis Rheum.* **44:** 2737–2745.

53. Kobelt, G., P. Lindgren, A. Singh & L. Klareskog. 2005. Cost effectiveness of etanercept (Enbrel) in combination with methotrexate in the treatment of active rheumatoid arthritis based on the TEMPO trial. *Ann. Rheum. Dis.* **64:** 1174–1179.

54. Genovese, M.C. 2009. Inhibition of p38: has the fat lady sung? *Arthritis Rheum.* **60:** 317–320.

55. Cohen, S.B., T.-T. Cheng, V. Chindalore, *et al.* 2009. Evaluation of the efficacy and safety of pamapimod, a p38 MAP kinase inhibitor, in a double-blind, methotrexate-controlled study of patients with active rheumatoid arthritis. *Arthritis Rheum.* **60:** 335–344.

56. Weinblatt, M.E., A. Kavanaugh, R. Burgos-Vargas, *et al.* 2008. Treatment of rheumatoid arthritis with a Syk kinase inhibitor. *Arthritis Rheum.* **58:** 3309–3318.

57. Smolen, J.S., A. Beaulieu, A. Rubbert-Roth *et al.* : OPTION Investigators. 2008. Effect of interleukin-6 receptor inhibition with tocilizumab in patients with rheumatoid arthritis (OPTION study): a double-blind, placebo-controlled, randomized trial. *Lancet* **371:** 987–997.

58. Keystone, E.C., M.C. Genovese, L. Klareskog *et al.* 2008. Golimumab, a human antibody to TNF-α given by monthly subcutaneous injections, in active rheumatoid arthritis despite methotrexate: the GO-FORWARD Study. *ARD Online*, December.

59. Bender, N.K., C.E. Heilig, B. Droll, *et al.* 2007. Immunogenicity, efficacy and adverse events of adalimumab in RA patients. *Rheumatol. Int.* **27:** 269–274.

60. Fefferman, D.S. & R.J. Farrell. 2005. Immunogenicity of biological agents in inflammatory bowel disease. *Inflamm. Bowel Dis.* **11:** 497–503.

61. Anderson, P.J. 2005. Tumor necrosis factor inhibitors: Clinical implications of their different immunogenicity profiles. *Semin. Arthritis Rheum.* **34:** 19–22.

62. Anderson, P., J. Louie, A. Lau & M. Broder. 2005. Mechanisms of differential immunogenicity of tumor necrosis factor inhibitors. *Curr. Rheumatol. Rep.* **7:** 3–9.

63. Baert, F., M. Noman, S. Vermeire, *et al.* 2009. Influence of immunogenicity on the long-term efficacy of infliximab in Crohn's disease. *N. Engl. J. Med.* **348:** 601–608.

64. Aarden, L., S.R. Ruuls & G. Wolbink. 2008. Immunogenicity of anti-tumor necrosis factor antibodies – toward improved methods of anti-antibody measurement. *Curr. Opin. Immunol.* **20:** 431–435.

65. Wolbink, G.J., M. Vis, W. Lems, *et al.* 2006. Development of antiinfliximab antibodies and relationship to clinical response in patients with rheumatoid arthritis. *Arthritis Rheum.* **54:** 711–715.

66. Saag, K.G., G.G. Ten, N.M. Patkar, *et al.* 2008 American College of Rheumatology recommendations for the use of nonbiologic and biologic disease-modifying antirheumatic drugs in rheumatoid arthritis. *Arthritis Rheum.* **56:** 762–784.

67. Danila, M.I., N.M. Patkar, J.R. Curtis, *et al.* 2008. Biologics and heart failure in rheumatoid arthritis: are we any wiser? *Curr. Opin. Rheumatol.* **20:** 327–333.

68. Patkar, N.M., G.G. Teng, J.R. Curtis & K.G. Saag. 2008. Association of infections and tuberculosis with antitumor necrosis factor alpha therapy. *Curr. Opin. Rheumatol.* **20:** 320–326.

69. Curtis, J.R., J. Xi, N. Patkar, *et al.* 2007. Drug-specific and time-dependent risk of bacterial infection among patients with rheumatoid arthritis who were exposed to tumor necrosis factor alpha antagonists. *Arthritis Rheum.* **56:** 4226–4227.

70. Love, T. & D.H. Solomon. 2008. The relationship between cancer and rheumatoid arthritis: still a large research agenda. *Arthritis Rheum.* **10:** 109.

71. Setoguchi, S., S. Schneeweiss, J. Avorn, *et al.* 2008. Tumor necrosis factor-alpha antagonist use and heart failure in elderly patients with rheumatoid arthritis. *Am. Heart J.* **156:** 336–341.

72. Schneeweiss, S., S. Setoguchi, M.E. Weinblatt, *et al.* 2007. Anti-tumor necrosis factor alpha therapy and the risk of serious bacterial infections in elderly patients with rheumatoid arthritis. *Arthritis Rheum.* **56:** 1754–1764.

73. Emery, P., F.C. Breedveld, S. Hall, *et al.* 2008. Comparison of methotrexate monotherapy with a combination of methotrexate and etanercept in active, early, moderate to severe rheumatoid arthritis (COMET): a randomized, double-blind, parallel treatment trial. *Lancet* **372:** 375–382.

74. Goekoop-Ruiterman, Y.P., J.K. de Vries-Bouwstra, C.F. Allaart, *et al.* 2007. Comparison of treatment strategies in early rheumatoid arthritis: a randomized trial. *Ann. Intern. Med.* **146:** 406–415.

75. Solomon, D.H., M. Lunt & S. Schneeweiss. 2008. The risk of infection associated with tumor necrosis factor α antagonists. *Arthritis Rheum.* **58:** 919–928.

# Ustekinumab: Lessons Learned from Targeting Interleukin-12/23p40 in Immune-Mediated Diseases

Michael Elliott,[a] Jacqueline Benson,[b] Marion Blank,[b] Carrie Brodmerkel,[b] Daniel Baker,[b] Kristin Ruley Sharples,[c] and Philippe Szapary[b]

[a]TransForm Pharmaceuticals, Inc., Lexington, Massachusetts, USA

[b]Centocor Research and Development, Inc., Malvern, Pennsylvania, USA

[c]Centocor Ortho Biotech, Inc., Malvern, Pennsylvania, USA

Interleukin (IL)-12 and IL-23 are related cytokines that have been implicated in the pathogenesis of several immune-mediated disorders. IL-12 and IL-23 are heterodimers made up of a common p40 subunit complexed to unique p35 (IL-12) or p19 (IL-23) subunits. Ustekinumab is a human monoclonal antibody that specifically binds the p40 subunit of IL-12/23. Ustekinumab prevents IL-12 and IL-23 from binding their cell surface receptor complexes, thereby blocking the T helper (Th) 1 (IL-12) and Th17 (IL-23) inflammatory pathways. Here, we discuss the preclinical and human translational data supporting a role for IL-12/23 in the pathogenesis of immune-mediated disorders, and how that rationale was challenged in the clinic during the course of the ustekinumab development program in several indications including psoriasis, psoriatic arthritis, Crohn's disease, and multiple sclerosis. We review the key efficacy and safety data in each of these immune-mediated diseases and compare and contrast the safety lessons learned from IL-12/23 genetically-deficient mice and humans in context of the overall clinical trial experience with ustekinumab.

*Key words:* ustekinumab; interleukin-12; interleukin-23; interleukin-12/23p40; psoriasis; psoriatic arthritis; Crohn's disease; multiple sclerosis

## Introduction

The interleukin (IL)-12 family of cytokines, which includes IL-12 and IL-23, is known to play a role in regulating T cell immune responses.[1] IL-12 is a heterodimer consisting of two subunits, p35 and p40, linked by a disulfide bond. Antigen-producing cells express IL-12, which participates in cell-mediated immunity by binding to a dichain receptor complex expressed on the surface of T cells or natural killer cells. It is believed that the p40 subunit of IL-12 binds the IL-12 receptor beta 1 (IL-12Rβ1) receptor and the p35 subunit binds to the second receptor chain (IL-12Rβ2), resulting in intracellular signaling.[2] In CD4$^+$ T cells, IL-12 signaling, along with antigen presentation, is believed to shift cell differentiation toward the T helper (Th) 1 phenotype, and is associated with robust production of the proinflammatory cytokine, interferon gamma (IFN-γ).[3] Activation of the Th1 pathway is believed to promote immunity against some intracellular pathogens; however, IL-12 may also be abnormally regulated in many immune-mediated diseases, including psoriatic disorders, Crohn's disease, and multiple sclerosis.[4]

IL-23 is also a heterodimer, consisting of the IL-12p40 protein subunit covalently linked to

Address for correspondence: Michael Elliott, M.B., B.S., Ph.D., TransForm Pharmaceuticals, Inc., 29 Hartwell Avenue, Lexington, MA 02421. Voice: 781-674-7912; fax: 781-863-6519. MElliot2@its.jnj.com

Cytokine Therapies: Ann. N.Y. Acad. Sci. 1182: 97–110 (2009).
doi: 10.1111/j.1749-6632.2009.05070.x © 2009 New York Academy of Sciences.

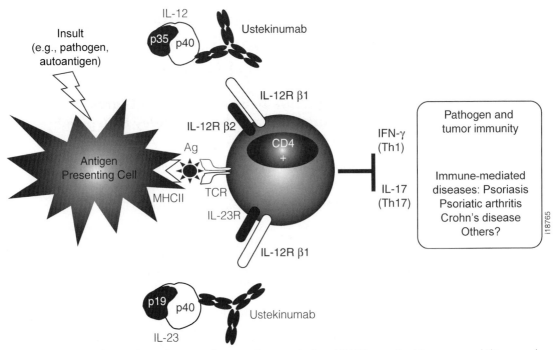

**Figure 1.** Ustekinumab mechanism of action. IL = interleukin; MHCII = major histocompatability complex II; Ag = antigen; TCR = T cell receptor; IL-23R = IL-23 receptor; IL-12R β1 and β2 = IL-12 receptor beta 1 and beta 2; IFN-γ = interferon-gamma; Th1 = T helper 1; Th17 = T helper 17.

a p19 partner protein. Consistent with IL-12 signaling, the p40 subunit binds to cell surface IL-12Rβ1; however, the IL-23p19 subunit binds to the second component of the IL-23 receptor complex, IL-23R, to induce IL-23-specific signaling.[5] Recently, IL-23 has been implicated in the maintenance of human Th17 cells, which can produce IL-17A, IL-17F, IL-22, IL-26, IFN-γ, CCL20,[6] and tumor necrosis factor-alpha (TNF-α).[7] IL-23 and the Th17 pathway may also contribute to the pathophysiology of several immune-mediated disorders.[7] Due to the commonality of p40 in both IL-12 and IL-23, p40 is a reasonable therapeutic target in certain disorders characterized by dysregulated CD4+ T cell function.

Ustekinumab (CNTO 1275; Stelara™; Centocor, Inc., Malvern, PA) is a human, immunoglobulin G1 kappa (IgG1κ) monoclonal antibody that specifically binds the shared p40 subunit of IL-12 and IL-23 and inhibits the interactions of IL-12 and IL-23 with the cell surface IL-12Rβ1 receptor (Fig. 1). Thus, the binding of ustekinumab prevents IL-12- or IL-23-mediated signaling cascades. Ustekinumab is an excellent platform for studying the effects of simultaneous interruption of Th1 and Th17 pathways in a variety of immune-mediated diseases. Here we review the preclinical rationale and published clinical trial results of ustekinumab treatment in psoriasis, psoriatic arthritis, Crohn's disease, and multiple sclerosis, as well as the human safety experience to-date of inhibition of the IL-12/23 pathway, contrasting it with what would be expected from IL-12/23 genetically-deficient mice and humans.

## Lessons Learned from Clinical Trials of Ustekinumab

Below, we summarize the published safety and efficacy data of ustekinumab across several diseases. For additional details, please refer to the cited primary publications.

## Psoriasis

Psoriasis is the most common chronic inflammatory skin disorder, affecting approximately 2 to 3% of the world's population.[8] Aberrant immune responses have been associated with the pathogenesis of psoriasis.[9–11] In particular, IL-12[12] and IL-23[13] have been implicated in the development of Th1 and Th17 immune responses in psoriasis, respectively. Preclinical studies using a murine model of psoriasis demonstrated that the injection of IL-12[14] or IL-23[15] into mice resulted in keratinocyte proliferation and histological alterations to the dermis, including the development of psoriatic plaques. In addition, psoriasis did not develop when an anti-IL-12 monoclonal antibody was administered.[14] Human psoriasis has been linked to genetic polymorphisms of the IL-12B gene, which encodes the p40 subunit of IL-12 and IL-23, and the IL-23R gene, which encodes the IL-23R receptor subunit.[16] The p40 subunit of IL-12[12] and IL-23[17] is overexpressed in psoriasis plaques, and studies have demonstrated increased IL-23p19 gene expression in psoriatic lesions.[15] Although the relative role of IL-12 and IL-23 in psoriasis requires further evaluation, IL-12/23p40 has been described as the "master switch" for psoriasis immunopathogenesis.[18]

A total of five clinical studies have been published to date to assess the efficacy and safety of ustekinumab in the treatment of moderate-to-severe plaque psoriasis, the first of several immune-mediated diseases that may potentially benefit from targeting IL-12/23p40. The proof-of-concept for ustekinumab for the treatment of moderate-to-severe psoriasis was evaluated in two Phase I studies.[19,20] These studies demonstrated significant and rapid clinical efficacy (as assessed by the Psoriasis Area and Severity Index [PASI] and Physician's Global Assessment [PGA]) in patients after receiving a single dose of ustekinumab. In addition, gene expression analyses of pre- and posttreatment lesional skin indicate that the administration of anti-IL-12/23p40 rapidly (i.e., by week 2)

downregulates mRNA expression of type 1 cytokines, as well as IL-12/23p40 and IL-23p19 (Fig. 2).[21] IL-12p35 was upregulated following anti-IL-12p40 administration, although IL-12p35 copy number was low and results were variable. In addition, IL-12p35 is differentially regulated as compared with IL-12p40 or IL-23p19 mRNA expression and may account for the differences between these molecules in response to ustekinumab treatment.[22] The rapid decrease of IL-12/23p40 and IL-23p19 expression levels preceded clinical response and histological changes.[21]

In a Phase II, double-blind, placebo-controlled study, 320 patients with moderate-to-severe psoriasis were randomized to receive ustekinumab (one 45-mg dose, one 90-mg dose, four weekly 45-mg doses, or four weekly 90-mg doses) or placebo.[23] The study design required patients in the ustekinumab group who did not respond (PGA ≥3) by week 16 to receive an additional dose, and those in the placebo group to receive a single dose of ustekinumab 90 mg at week 20. At week 12, 51.6% and 59.4% of patients who received ustekinumab 45 mg and 90 mg, respectively, and 67.2% and 81.3% who received four weekly 45-mg and 90-mg doses, respectively, achieved at least a 75% improvement from baseline in PASI (PASI 75), compared with 1.6% of patients who received placebo ($P < 0.001$ each).

Psoriatic lesions are characterized by thickened epidermal layers, which result from excessive keratinocyte cell proliferation. Disease resolution at a histological level is characterized by a thinning of the epidermal layer and an absence of rete (epidermal) ridges. Histological analyses were completed in the Phase II study at baseline and week 12 to understand the effects of ustekinumab on standard hallmarks of psoriasis. Consistent with the observed clinical benefit, ustekinumab significantly reduced the epidermal thickness of lesional skin (Fig. 3). This study confirmed the efficacy of ustekinumab (clinically and histologically) in the treatment of moderate-to-severe psoriasis and helped determine the appropriate dosage and dosing

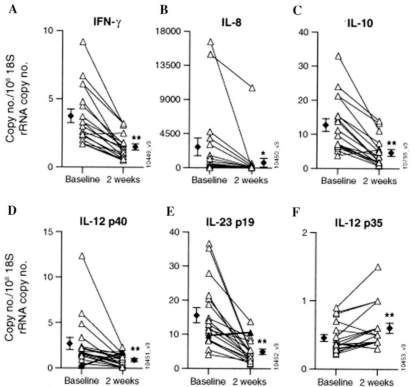

**Figure 2.** mRNA reduction of multiple cytokines including p40 and p19 by week 2 in patients with psoriasis treated with ustekinumab. The mean ± standard error of all patients is indicated by the diamond symbol (♦). IFN-γ = interferon-gamma; IL = interleukin; * = $P < 0.05$; ** = $P < 0.01$; statistically significant differences between baseline and 2 weeks posttreatment. Adapted with permission from Toichi, E., Torres, G., McCormick, T.S., et al. 2006. An anti-Il-12p40 antibody down-regulates type 1 cytokines, chemokines, and Il-12/IL-23 in psoriasis. *J. Immunol.* **177:** 4917–4926. © 2006. The American Association of Immunologists, Inc.

interval. Additional studies were conducted to confirm the efficacy and safety of ustekinumab in a larger patient population and for a longer duration.

Two Phase III studies, PHOENIX 1[24] and PHOENIX 2,[25] evaluated the efficacy of two dosing regimens of ustekinumab (45 mg and 90 mg) for the treatment of moderate-to-severe psoriasis. PHOENIX 1 and 2 were randomized, double-blind, placebo-controlled studies in which a total of 1,996 patients were randomized to receive ustekinumab 45 or 90 mg at weeks 0 and 4, and followed by every 12 weeks thereafter, or placebo. At week 12, patients receiving placebo crossed over to receive ustekinumab 45 mg or 90 mg every 12 weeks.

These studies included a placebo-controlled period (weeks 0 to 12) and a placebo-crossover and active treatment period (PHOENIX 1: weeks 12 to 40; PHOENIX 2: weeks 12 to 28). Thereafter, patients entered a randomized withdrawal period (weeks 40 to 76) in PHOENIX 1, and a dose optimization period (weeks 28 to 52) in PHOENIX 2.

In these two studies,[24,25] a rapid onset of efficacy was observed following ustekinumab treatment, with approximately 10% of patients in the 45 mg and 90 mg groups achieving a PASI 50 response at week 2. By week 4, a significantly greater proportion of patients in the ustekinumab 45 mg and 90 mg groups achieved a PASI 75 response compared with the placebo

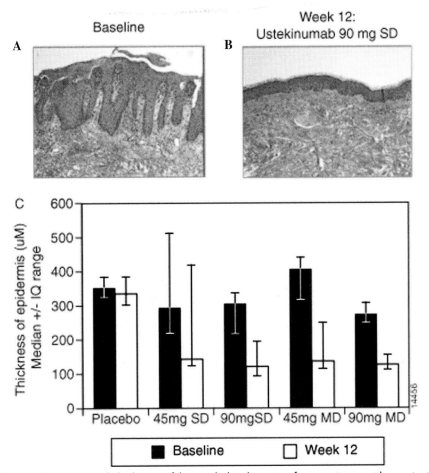

**Figure 3.** Epidermal thickness of lesional skin biopsies from patients with psoriasis at baseline and at week 12 in patients treated with ustekinumab. SD = single dose; MD = multiple doses; IQ = interquartile.

group ($P < 0.001$), after receiving only one dose of ustekinumab. After one dose, higher proportions of patients in the ustekinumab group (66.4% to 75.7%) achieved the primary endpoint (PASI 75 response at week 12; Fig. 4), than in the placebo group, a finding that was consistent with the Phase II results. Following ustekinumab maintenance dosing, clinical response rates continued to improve, and peak responses were observed from weeks 20 to 24.

In the PHOENIX 1 trial,[24] the benefits of long-term maintenance therapy were evaluated in patients with a PASI 75 response at weeks 28 and 40. Patients were randomized at week 40 to receive maintenance therapy or withdrawal of therapy. A significantly higher proportion of patients who continued ustekinumab therapy maintained PASI 75 response through at least one year compared with patients withdrawn from treatment ($P < 0.001$). In fact, psoriasis returned in patients after missing one dose of ustekinumab. No rebound of psoriasis was observed with withdrawal of treatment; however, 85.6% of patients withdrawn from therapy, and who lost therapeutic effect, regained PASI 75 response within 12 weeks of reinitiation of therapy.

Overall, the Phase II and III studies consistently demonstrated that ustekinumab was highly efficacious and led to rapid and significant improvements in the signs and symptoms of moderate-to-severe psoriasis. The

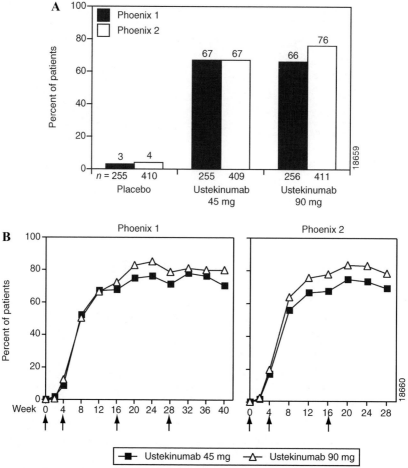

**Figure 4.** Proportion of patients in the PHOENIX 1[24] and PHOENIX 2[25] psoriasis clinical trials achieving at least a 75% improvement from baseline in Psoriasis Area and Severity Index (PASI 75) response at **A)** week 12 and **B)** through weeks 28/40. Arrows indicate visits at which ustekinumab was administered.

proportion of patients who responded to ustekinumab treatment (assessed by PASI 75 and PGA of cleared or minimal) increased over time, reaching maximal or near maximal response rates approximately 6 months after initiating therapy. At week 28 (after three injections), more than 90% of ustekinumab-treated patients reported a clinically meaningful improvement in psoriasis (PASI 50 response). Between 59% and 70% of patients achieved a PGA of cleared or minimal, confirming a high level of efficacy. Ustekinumab was effective across all demographics (age, sex, and race), and efficacy was not affected by baseline disease severity or other disease characteristics.

The strength and consistency of these findings notwithstanding, ustekinumab is not a cure for psoriasis.

Our current understanding of the safety of ustekinumab in psoriasis is derived from 2,266 patients treated through 18 months (2,251 patient years of exposure). Safety comparisons are reviewed during the placebo-controlled portion, as well as through the controlled and uncontrolled period with up to 18 months of follow-up. Ustekinumab was generally well-tolerated, and rates of adverse events (AEs) were comparable between patients receiving ustekinumab (54.6%) or placebo (50.4%) through the common 12-week placebo-controlled period of the studies. The most commonly reported AEs were nasopharyngitis (8.2% and 7.9%, respectively) and upper respiratory tract infection (5.4% and 4.4%, respectively). Rates of serious adverse events (SAEs) through week 12 were also comparable between patients in the ustekinumab group (1.5%) and the placebo group (1.4%). The incidences of serious infections (1.2 and 1.7 per 100 patient-years of follow-up, respectively) and malignancies other than nonmelanoma skin cancer (NMSC) (0.25 and 0.57 per 100 patient-years, respectively) in ustekinumab-treated patients were low and comparable with placebo-treated patients through the controlled portion of the studies. Through 18 months of follow-up, rates of serious infections (1.1 per 100 patient-years of follow-up) and non-NMSC malignancies (0.36 per 100 patient-years of follow-up) remained low and were stable in the combined ustekinumab group. Through up to one year of follow-up in the combined Phase II and Phase III studies, rates of injections with injection-site reactions in ustekinumab-treated patients were low (45 mg: 1.0%; 90 mg: 1.3%) and similar to those in placebo-treated patients (0.4%). Additionally, antibodies to ustekinumab in the Phase III studies were low (5.4%) through 18 months of follow-up for the combined ustekinumab group. Despite the population size and duration of these studies, the potential exists for the development of an unexpected event(s) to occur following a longer duration of exposure, or in larger patient populations. Furthermore, additional data will be collected in ongoing long-term extensions with up to five years of treatment. Overall, ustekinumab has shown excellent efficacy in psoriasis, suggesting a central role of Th1/Th17 pathways in this disease, with a favorable risk-benefit profile through up to 1.5 years of treatment.

## Psoriatic Arthritis

Psoriatic arthritis (PsA) is a chronic inflammatory arthropathy occurring in approximately 10% to 25% of patients with psori-

asis.[26,27] Animal models[28,29] and clinical observations[30,31] suggest that IL-12 and IL-23 contribute to the manifestations of symptoms and joint changes in immune-mediated arthritis. Indeed, IL-12 and IL-23 have been shown to mediate collagen-induced arthritis, a widely accepted model of polyarticular arthritis.[32] Compared with wild-type mice, mice deficient in IL-12 and IL-23, or IL-23 alone, are protected from arthritis when immunized with collagen.[28,33]

While disease-modifying antirheumatic drugs (DMARDs) and biologics targeting TNF-$\alpha$ are effective in the treatment of PsA,[34–36] some patients with PsA are not completely responsive to, or are intolerant of, these treatments. As such, the efficacy and safety of ustekinumab in PsA was evaluated in a Phase II, proof-of-concept, multicenter, double-blind, placebo-controlled study.[27]

Patients with active PsA were randomized (1:1) to receive subcutaneous injections of either ustekinumab 90 mg at weeks 0, 1, 2, and 3 (total dose 360 mg), followed by placebo at weeks 12 and 16 (Group 1), or placebo at weeks 0, 1, 2, and 3, followed by ustekinumab 90 mg at weeks 12 and 16 (total dose 180 mg; Group 2).[27] The primary efficacy analysis was a comparison between Groups 1 and 2 of the proportions of patients achieving a 20% improvement from baseline in the American College of Rheumatology (ACR20) criteria at week 12. At week 12, a greater proportion of patients in Group 1 (42.1%) achieved an ACR20 response compared with Group 2 (14.3%; $P = 0.0002$). Among the 85% of patients with $\geq$ 3% of body surface area involvement, a significantly greater proportion of patients achieved a PASI 75 response at week 12 in Group 1 (52.4%) than in Group 2 (5.5%; $P < 0.0001$).[27]

This Phase II study[27] demonstrated that ustekinumab significantly reduces the signs and symptoms of PsA and improves skin lesions compared with placebo. Patients receiving two doses of ustekinumab following placebo-crossover had similar response rates to those initially randomized to ustekinumab. The efficacy of two or four doses of ustekinumab was comparable with that of other biologic therapies for PsA. This proof-of-concept study in PsA suggests a role of IL-12/23p40 in joint disease and was the foundation for the planned Phase III clinical program.

The safety profile of ustekinumab in patients with PsA was generally consistent with that observed in patients with psoriasis. In the Phase II study,[27] the proportion of patients who reported at least one AE through week 12 (placebo-controlled period) was comparable between patients receiving four weekly doses of ustekinumab (60.5%) or placebo (62.9%). The most commonly reported AEs in both Group 1 and Group 2 were upper respiratory tract infection (13.2% and 8.6%, respectively), nasopharyngitis (10.5% and 2.9%, respectively), diarrhea (6.6% and 2.9%, respectively), and headache (6.6% and 4.3%, respectively). Through week 12, no ustekinumab-treated patients and three placebo-treated patients reported SAEs. In addition, no substantial differences in safety were observed between patients who received the two-dose or four-dose regimen of ustekinumab. Overall, ustekinumab significantly reduced the signs and symptoms of PsA, with a safety profile comparable to that observed in psoriasis.

## Crohn's Disease

In addition to dermatologic disorders, IL-12 and IL-23 have also been implicated in gastrointestinal disorders. Crohn's disease is an inflammatory disease of the gastrointestinal tract with an average annual incidence rate of 6.3 per 100,000 people in the United States (US).[37] The role of IL-12 in Crohn's disease was demonstrated in murine models of the disease in which anti-IL-12p40 antibodies administered at early or late time points improved clinical and histopathological changes.[38,39] Similarly, the role of IL-23 in Crohn's disease was demonstrated in a T-cell-deficient murine model of chronic intestinal inflammation,[40] as well as a study associating the IL-23R gene and

Crohn's disease.[41] The colitis was suppressed in recipient mice deficient in p19 and p40 but not in p35, the subunit specific for IL-12.[42,43] The role of IL-12/23p40 in Crohn's disease in humans was supported by a small study reporting efficacy of an anti-IL-12/23p40 antibody in patients with Crohn's disease.[44] Considering the improvements of psoriasis with ustekinumab treatment, and the preclinical and clinical studies of IL-12/23p40 inhibitors in Crohn's disease, the efficacy and safety of ustekinumab for the treatment of Crohn's disease was evaluated.[45]

An early Phase IIa study examined the clinical effects of ustekinumab in patients with moderate-to-severe Crohn's disease.[45] In this double-blind study, 104 patients failing conventional or biologic therapy (Population 1) were randomized to receive: subcutaneous (SC) placebo at weeks 0, 1, 2, and 3, followed by ustekinumab 90 mg at weeks 8, 9, 10, and 11; SC ustekinumab at weeks 0, 1, 2, and 3, followed by placebo at weeks 8, 9, 10, and 11; intravenous (IV) placebo at week 0, followed by ustekinumab 4.5 mg/kg at week 8; or IV ustekinumab 4.5 mg/kg at week 0, followed by placebo at week 8. Additionally, 27 primary or secondary nonresponders to the anti-TNF agent, infliximab (Population 2), were randomized to receive open-label SC ustekinumab at weeks 0, 1, 2, and 3, or IV ustekinumab 4.5 mg/kg at week 0. Clinical response was defined as a reduction of at least 25% and 70 points in the Crohn's disease activity index score from baseline. Patients were followed for safety and efficacy through week 28.

A greater proportion of ustekinumab-treated patients in Population 1 achieved clinical response than placebo-treated patients at weeks 4 and 6 (52.9% and 30.2%, respectively, at both time points; $P = 0.02$) and at the week 8, the primary end point (49.0% and 39.6%, respectively; $P = 0.34$). In a subgroup of 49 patients in Population 1 who had previously received infliximab (neither primary nor secondary nonresponders), clinical response rates at weeks 2, 4, 6, and 8 were significantly greater among ustekinumab-treated patients than placebo-treated patients (55% to 59% vs. 15% to 26%, respectively; $P < 0.05$). Median C-reactive protein concentrations were unchanged or increased at week 8 in patients who received placebo, whereas these values were decreased at week 8 in Population 1 patients who received ustekinumab. At week 8, following receipt of SC or IV ustekinumab, clinical response rates among Population 2 patients were 42.9% and 53.8%, respectively, which were similar to those observed in Population 1. Additionally, median C-reactive protein concentrations were decreased at week 8 in Population 2 patients who received IV or SC ustekinumab.

Although the design of this study precludes clear conclusions about efficacy, this study demonstrated that ustekinumab may induce clinical response in patients with moderate-to-severe Crohn's disease. Treatment effects were greatest at weeks 4 to 6, and in patients who had previously been treated with infliximab, making ustekinumab a potential therapeutic alternative for this patient population. The relative importance of interrupting IL-12/23 will be elucidated by a larger, ongoing Phase IIb study.

Treatment with ustekinumab was also generally well tolerated in patients with Crohn's disease.[45] Through the placebo-controlled period (week 8), a similar frequency of AEs was observed in Population 1 patients who received ustekinumab (71.2%) compared with placebo (78.8%). Gastrointestinal disorders were the most commonly reported AEs (32.7% in the combined IV and SC ustekinumab group versus 48.1% in the combined IV and SC placebo group) (data on file, Centocor, Inc.). Through week 8, two ustekinumab-treated patients (3.8%) and three placebo-treated patients (5.8%) reported at least one SAE. A total of nine SAEs were reported, of which seven were related to Crohn's disease. The incidence of infections was similar between patients treated with ustekinumab and placebo. A case of disseminated histoplasmosis occurred in a patient who had discontinued infliximab 3 months prior to study entry, was febrile at study baseline, and received IV ustekinumab, azathioprine, and prednisone concomitantly during the study. Overall, a favorable clinical response was observed in patients treated with ustekinumab, a response that will be explored more fully in a larger ongoing study.

## Multiple Sclerosis

Multiple sclerosis (MS) is a neurological disorder affecting the central nervous system (CNS). The estimated prevalence of MS in the US population is 85 per 100,000 people.[46] Studies supporting the role of IL-12[47] and IL-23[48] in MS have been performed in an animal model of demyelination (experimental autoimmune encephalitis [EAE] model) that is similar to human MS. IL-12 can exacerbate EAE,[49] whereas treatment with neutralizing anti-IL-12p40 antibodies has been shown to delay the onset or prevent relapses of EAE in both murine models[49,50] and primates.[51] Likewise, the role of IL-23 in EAE has been demonstrated using mice deficient in IL-12 or IL-23 subunits. A recent study using IL-23p19-deficient mice demonstrated complete resistance to disease induction, similar to IL-12p40-deficient mice.[52] IL-12/23p40[53] and IL-23p19[54] have also been detected in human MS lesions, and elevated levels of IL-12 have been observed in the CNS of patients with MS.[55] These findings from preclinical studies suggested the potential therapeutic benefit of anti-IL-12/23 antibodies in human MS.

A small ($n = 20$) Phase I study[56] and a larger Phase II, multicenter, randomized, double-blind, placebo-controlled study[57] were performed to evaluate the efficacy of ustekinumab in patients with relapsing-remitting MS. In the Phase II study,[57] 249 patients were randomized (1:1:1:1:1) to receive SC injections of placebo, or ustekinumab 27 mg, 90 mg, or 180 mg (the highest dose of ustekinumab administered in any study to-date) every 4 weeks, or ustekinumab 90 mg every 8 weeks (through week 17), following an induction phase during which all patients received placebo or ustekinumab weekly from baseline to week 3. Through week 23, there were no significant or clinically meaningful differences in the cumulative number of gadolinium-enhancing T1-weighted lesions on serial cranial magnetic resonance imaging (the primary end point) between the ustekinumab and placebo groups. The cumulative number of new lesions was similar in all groups, and the proportion of patients with at least one new lesion was comparable between the combined ustekinumab (64.0%) and placebo (61.2%) groups.[57] Overall, ustekinumab was not shown to be efficacious for the treatment of MS.

In patients with MS, ustekinumab was generally safe and well tolerated and did not appear to exacerbate or inhibit inflammatory demyelination.[57] As a result of the lack of demonstrated efficacy in this patient population, the long-term safety follow-up was terminated at week 37. AEs occurred in 85.0% of ustekinumab-treated patients compared with 77.6% of placebo-treated patients. Rates of AEs that were higher in ustekinumab-treated patients than in placebo-treated patients included: injection-site erythema (22.5% vs. 8.2%), headache (14.0% vs. 4.1%), and fatigue (13.0% vs. 2.0%), respectively. The proportion of patients reporting SAEs was generally low and comparable between patients receiving ustekinumab (3.0%) and placebo (2.0%). No serious infections were reported and no difference was observed in the proportion of patients with infections in the combined ustekinumab treatment group compared with the placebo group. Overall, despite the lack of efficacy in MS, the disease did not appear to worsen following ustekinumab treatment.

## Discussion

In total, 2,770 patients have received ustekinumab across nine clinical trials for the treatment of various immune-mediated disorders including psoriasis ($n = 2,301$), PsA ($n = 133$),

Crohn's disease ($n = 120$), and MS ($n = 216$). Ustekinumab was shown to be efficacious in the treatment of psoriasis and PsA. While ustekinumab was clearly not efficacious for MS, its effect in Crohn's disease is being evaluated in a larger ongoing trial.

It is noteworthy that whereas all four diseases studied had strong preclinical animal model and human translational rationale for the targeting of IL-12/23p40, the demonstration of clinical efficacy was convincing in psoriasis and PsA, and suggestive in Crohn's disease. MS is the exception and deserves comment. At least four hypotheses for this failure can be advanced. First is the notion that IL-23 may be the major T cell driver in this disease, and that dual IL-12/23 inhibition may be less efficient than inhibition of IL-23 alone as a result of cross-regulation between the Th1 and Th17 pathways.[58] Second is the possibility that neither cytokine is important in the disease, although this seems unlikely given the wealth of preclinical and human translational rationale. Third is that the timing of administration[59] or duration of observation in the Phase II trial was insufficient to establish benefit from inhibition of an upstream driver, although this is clearly not the case in other immune-mediated diseases, where ustekinumab is effective within weeks. Finally and perhaps more likely than any of the previous hypotheses, it is probable that ustekinumab, with an approximate molecular weight of 150 kDa, did not cross the blood-brain barrier to any significant degree, and therefore did not reach the site of action of IL-12/23 in human MS. If correct, this would suggest the conclusion that IL-12/23 are unimportant in the peripheral (non-CNS) component of established MS; however, the possibility remains that they are indeed of importance in the brain. This final hypothesis could be tested with an appropriate, possibly low molecular weight inhibitor.

As with other immunomodulating agents, the long-term use of ustekinumab raises theoretical concerns for infections and malignancies. This potential risk is supported by data from studies in mice, suggesting that IL-12 (Th1) and IL-23 (Th17) responses may contribute to protective immune responses to viral and bacterial pathogens.[60,61] These observations are primarily reported in genetically deficient "knockout" mice or following very high doses of neutralizing antibody administration. There are also human IL-12/23-deficient patients who may shed further light on the role of IL-12 and IL-23 in pathogen immunity. Humans with genetic deficiencies in the Th1 and Th17 pathways, including IL-12/23p40, IL-12Rβ1, IFN-γ receptor 1 and 2, signal transducer and activator of transcription 1, and nuclear factor-κB essential modulator have been reported.[61] In contrast to observations in murine models, these individuals have normal resistance to infections;[61,62] however, they are susceptible to non-tuberculosis (TB) primary *Mycobacteria* infection and recurring *Salmonella spp.*[61] This is in contrast to what was observed in the clinical program of ustekinumab, where no cases of typical or atypical mycobacterial or salmonella infections were reported. The reasons for the discrepancy between the genetically-deficient patients (with a clear risk of mycobacterial infection) and the ustekinumab-treated patients (with no demonstrated risk to date) are likely to be many, and may include that the clinical trial patients were screened for latent TB at baseline, with subsequent treatment of positive cases with isoniazid. Alternative or additional explanations include the limited exposure to date in large endemic populations and the likelihood that ustekinumab treatment does not confer complete inhibition of IL-12/23p40, as is observed with genetic deficiencies.

Similar to models of pathogen immunity, rodent tumor models suggest that mouse IL-12 can promote antitumor effects and that IL-12/IL-23p40-deficient mice, or mice treated with anti-IL-12/23p40 antibody, have decreased host defense to tumors. The relevance of these experimental findings in murine models for malignancy risk in humans is unknown. Of note, there have been no published reports of malignancy in IL-12/23-deficient humans,[63] although one case of esophageal carcinoma has been verbally communicated (J.L. Casanova, oral communication, May 2008). This cohort of patients is small ($n < 300$) and young ($<40$ years old). While the clinical trial data with ustekinumab through 18 months of therapy are reassuring with regard to malignancy risk, longer-term follow-up will be needed to assess the net effect of pharmacologic inhibition of IL-12/23p40 on the incidence of cancer.

Overall, ustekinumab was generally well tolerated in clinical studies across indications. The incidences of infections and malignancies in ustekinumab-treated patients were low. Of note, no cases of TB, atypical mycobacterial infection, salmonella infection, lymphoma, or demyelinating disorders were reported in any of these ustekinumab clinical trials through up to 1.5 years of treatment. It should be noted that these studies were conducted with different dosing regimens, frequencies of administration, treatment durations, and routes of administration. While the emerging safety profile has demonstrated a positive risk-benefit profile for ustekinumab, rare events and events that may occur during longer periods of follow-up cannot be ruled out and need to be closely observed during the ongoing long-term extensions of the Phase III psoriasis trials (up to 5 years), as well as large, planned observational studies.

## Summary

The development of ustekinumab in immune-mediated disease was justified by studies showing an important role for IL-12/23 p40 in preclinical disease models and by strong genetic association and other translational data in humans. Although systemic ustekinumab treatment was effective in ameliorating nonhuman primate EAE under both preventative and therapeutic dosing regimens, it proved to be ineffective in treating patients with relapsing remitting MS. The reasons for this are unclear, but likely include issues of poor access of the drug to the CNS. In psoriasis and PsA however, ustekinumab showed impressive efficacy in the clinic and thus provides clear target validation in these diseases. Also, in Crohn's disease, the efficacy observed in a Phase II study was sufficient to warrant further clinical investigation. Although preclinical and human genetic data suggested possible safety liabilities for IL-12/23 blockade, to date, the safety profile of ustekinumab in the clinic has been good. Taken together, the ustekinumab treatment data reviewed here validate the importance of IL-12/23p40 in human immune-mediated disease, and offer the promise of significant patient benefit.

## Acknowledgments

The authors wish to thank Mary Whitman, PhD, Michelle Perate, Rebecca Clemente, PhD, Mary Ann Thomas, RN, and James Barrett of Centocor Ortho Biotech, Inc. for their writing support.

## Conflicts of Interest

All authors are employees of subsidiaries of Johnson & Johnson.

## References

1. Lyakh, L., G. Trinchieri, L. Provezza, *et al*. 2008. Regulation of interleukin-12/interleukin-23 production and the T-helper 17 response in humans. *Immunol. Rev.* **226:** 112–131.
2. Presky, D.H., H. Yang, L.J. Minetti, *et al*. 1996. A functional interleukin 12 receptor complex is composed of two beta-type cytokine receptor subunits. *Proc. Natl. Acad. Sci. USA* **93:** 14002–14007.
3. Trinchieri, G., S. Pflanz, & R.A. Kastelein. 2003. The IL-12 family of heterodimeric cytokines: new players in the regulation of T cell responses. *Immunity* **19:** 641–644.
4. Boniface, K., B. Blom, Y.J. Liu, *et al*. 2008. From interleukin-23 to T-helper 17 cells: human T-helper cell differentiation revisited. *Immunol. Rev.* **226:** 132–146.
5. Parham, C., M. Chirica, J. Timans, *et al*. 2002. A receptor for the heterodimeric cytokine IL-23 is composed of IL-12Rβ1 and a novel cytokine receptor subunit, IL-23R. *J. Immunol.* **168:** 5699–5708.
6. Wilson, N.J., K. Boniface, J.R. Chan, *et al*. 2007. Development, cytokine profile and function of human interleukin 17-producing helper T cells. *Nat. Immunol.* **8:** 950–957.
7. Kikly, K., L. Liu, S. Na, *et al*. 2006. The IL-23/Th(17) axis: therapeutic targets for autoimmune inflammation. *Curr. Opin. Immunol.* **18:** 670–675.
8. Schön, M.P. & W.H. Boehncke. 2005. Psoriasis. *N. Engl. J. Med.* **352:** 1899–1912.
9. Gottlieb, A.B., R. Evans, S. Li, *et al*. 2004. Infliximab induction therapy for patients with severe plaque-type psoriasis: a randomized, double-blind, placebo-controlled trial. *J. Am. Acad. Dermatol.* **51:** 534–542.
10. Reich, K., F.O. Nestle, K. Papp, *et al*. 2005. Infliximab induction and maintenance therapy for moderate-to-severe psoriasis: a phase III, multicentre, double-blind trial. *Lancet* **366:** 1367–1374.
11. Leonardi, C.L., J.L. Powers, R.T. Matheson, *et al*. 2003. Etanercept as monotherapy in patients with psoriasis. *N. Engl. J. Med.* **349:** 2014–2022.
12. Yawalkar, N., S. Karlen, R. Hunger, *et al*. 1998. Expression of interleukin-12 is increased in psoriatic skin. *J. Invest. Dermatol.* **111:** 1053–1057.
13. Lee, E., W.L. Trepicchio, J.L. Oestreicher, *et al*. 2004. Increased expression of interleukin 23 p19 and p40 in lesional skin of patients with psoriasis vulgaris. *J. Exp. Med.* **199:** 125–130.
14. Hong, K., A. Chu, B.R. Lúdvíksson, *et al*. 1999. IL-12, independently of IFN-γ, plays a crucial role in the pathogenesis of a murine psoriasis-like skin disorder. *J. Immunol.* **162:** 7480–7491.
15. Chan, J.R., W. Blumenschein, E. Murphy, *et al*. 2006. IL-23 stimulates epidermal hyperplasia via TNF and IL-20R2-dependent mechanisms with implications for psoriasis pathogenesis. *J. Exp. Med.* **203:** 2577–2587.
16. Cargill, M., S.J. Schrodi, M. Chang, *et al*. 2007. A large-scale genetic association study confirms IL12B and leads to the identification of IL23R as psoriasis-risk genes. *Am. J. Hum. Genet.* **80:** 273–290.
17. Piskin, G., R.M.R. Sylva-Steenland, J.D. Bos, *et al*. 2006. In vitro and in situ expression of IL-23 by keratinocytes in healthy skin and psoriasis lesions: enhanced expression in psoriatic skin. *J. Immunol.* **176:** 1908–1915.
18. Nestle, F.O. & C. Conrad. 2004. The IL-12 family member p40 chain as a master switch and novel therapeutic target in psoriasis. *J. Invest. Dermatol.* **123:** xiv–xv.
19. Kauffman, C.L., N. Aria, E. Toichi, *et al*. 2004. A phase I study evaluating the safety, pharmacokinetics, and clinical response of a human IL-12 p40 antibody

in subjects with plaque psoriasis. *J. Invest. Dermatol.* **123:** 1037–1044.

20. Gottlieb, A.B., K.D. Cooper, T.S. McCormick, *et al.* 2007. A phase 1, double-blind, placebo-controlled study evaluating single subcutaneous administrations of a human interleukin-12/23 monoclonal antibody in subjects with plaque psoriasis. *Curr. Med. Res. Opin.* **23:** 1081–1092.

21. Toichi, E., G. Torres, T.S. McCormick, *et al.* 2006. An anti-IL-12p40 antibody down-regulates type 1 cytokines, chemokines, and IL-12/IL-23 in psoriasis. *J. Immunol.* **177:** 4917–4926.

22. Liu, J., S. Cao, L.M. Herman, *et al.* 2003. Differential regulation of interleukin (IL)-12 p35 and p40 gene expression and interferon (IFN)-γ-primed IL-12 production by IFN regulatory factor 1. *J. Exp. Med.* **198:** 1265–1276.

23. Krueger, G.G., R.G. Langley, C. Leonardi, *et al.*; CNTO 1275 Psoriasis Study Group. 2007. A human interleukin-12/23 monoclonal antibody for the treatment of psoriasis. *N. Engl. J. Med.* **356:** 580–592.

24. Leonardi, C.L., A.B. Kimball, K.A. Papp, *et al.*; PHOENIX 1 study investigators. 2008. Efficacy and safety of ustekinumab, a human interleukin-12/23 monoclonal antibody, in patients with psoriasis: 76-week results from a randomised, double-blind, placebo-controlled trial (PHOENIX 1). *Lancet* **371:** 1665–1674.

25. Papp, K.A., R.G. Langley, M. Lebwohl, *et al.*; PHOENIX 2 study investigators. 2008. Efficacy and safety of ustekinumab, a human interleukin-12/23 monoclonal antibody, in patients with psoriasis: 52-week results from a randomised, double-blind, placebo-controlled trial (PHOENIX 2). *Lancet* **371:** 1675–1684.

26. Liu, Y., C. Helms, W. Liao, *et al.* 2008. A genome-wide association study of psoriasis and psoriatic arthritis identifies new disease loci. *PLoS. Genet.* **4:** e1000041.

27. Gottlieb, A., A. Menter, A. Mendelsohn, *et al.* 2009. Ustekinumab, a human interleukin 12/23 monoclonal antibody, for psoriatic arthritis: randomised, double-blind, placebo-controlled, crossover trial. *Lancet* **373:** 633–640.

28. Murphy, C.A., C.L. Langrish, Y. Chen, *et al.* 2003. Divergent pro- and antiinflammatory roles for IL-23 and IL-12 in joint autoimmune inflammation. *J. Exp. Med.* **198:** 1951–1957.

29. Malfait, A.M., D.M. Butler, D.H. Presky, *et al.* 1998. Blockade of IL-12 during the induction of collagen-induced arthritis (CIA) markedly attenuates the severity of the arthritis. *Clin. Exp. Immunol.* **111:** 377–383.

30. Ribbens, C., B. André, O. Kaye, *et al.* 2000. Increased synovial fluid levels of interleukin-12, sTNF-RII/sTNF-RI ratio delineate a cytokine pattern

characteristic of immune arthropathies. *Eur. Cytokine. Netw.* **11:** 669–676.

31. Ritchlin, C., S.A. Haas-Smith, D. Hicks, *et al.* 1998. Patterns of cytokine production in psoriatic synovium. *J. Rheumatol.* **25:** 1544–1552.

32. Germann, T., J. Szeliga, H. Hess, *et al.* 1995. Administration of interleukin 12 in combination with type II collagen induces severe arthritis in DBA/1 mice. *Proc. Natl. Acad. Sci. USA* **92:** 4823–4827.

33. McIntyre, K.W., D.J. Shuster, K.M. Gillooly, *et al.* 1996. Reduced incidence and severity of collagen-induced arthritis in interleukin-12-deficient mice. *Eur. J. Immunol.* **26:** 2933–2938.

34. Antoni, C.E., A. Kavanaugh, B. Kirkham, *et al.* 2005. Sustained benefits of infliximab therapy for dermatologic and articular manifestations of psoriatic arthritis: results from the infliximab multinational psoriatic arthritis controlled trial (IMPACT). *Arthritis Rheum.* **52:** 1227–1236.

35. Mease, P.J., A.J. Kivitz, F.X. Burch, *et al.* 2004. Etanercept treatment of psoriatic arthritis: safety, efficacy, and effect on disease progression. *Arthritis Rheum.* **50:** 2264–2272.

36. Mease, P.J., D.D. Gladman, C.T. Ritchlin, *et al.* 2005. Adalimumab for the treatment of patients with moderately to severely active psoriatic arthritis: results of a double-blind, randomized, placebo-controlled trial. *Arthritis Rheum.* **52:** 3279–3289.

37. Herrinton, L.J., L. Liu, J.D. Lewis, *et al.* 2008. Incidence and prevalence of inflammatory bowel disease in a Northern California managed care organization, 1996–2002. *Am. J. Gastroenterol.* **103:** 1998–2006.

38. Neurath, M.F., I. Fuss, B.L. Kelsall, *et al.* 1995. Antibodies to interleukin 12 abrogate established experimental colitis in mice. *J. Exp. Med.* **182:** 1281–1290.

39. Davidson, N.J., S.A. Hudak, R.E. Lesley, *et al.* 1998. IL-12, but not IFN-γ, plays a major role in sustaining the chronic phase of colitis in IL-10-deficient mice. *J. Immunol.* **161:** 3143–3149.

40. Hue, S., P. Ahern, S. Buonocore, *et al.* 2006. Interleukin-23 drives innate and T cell-mediated intestinal inflammation. *J. Exp. Med.* **203:** 2473–2483.

41. Duerr, R.H., K.D. Taylor, S.R. Brant, *et al.* 2006. A genome-wide association study identifies IL23R as an inflammatory bowel disease gene. *Science* **314:** 1461–1463.

42. Neurath, M.F. 2007. IL-23: a master regulator in Crohn disease. *Nat. Med.* **13:** 26–28.

43. Peluso, I., F. Pallone, G. Monteleone. 2006. Interleukin-12 and Th1 immune response in Crohn's disease: pathogenetic relevance and therapeutic implication. *World J. Gastroenterol.* **12:** 5606–5610.

44. Mannon, P.J., I.J. Fuss, L. Mayer, *et al.*; Anti-IL-12 Crohn's Disease Study Group. 2004.

Anti-interleukin-12 antibody for active Crohn's disease. *N. Engl. J. Med.* **351:** 2069–2079.

45. Sandborn, W.J., B.G. Feagan, R.N. Fedorak, *et al.*; Ustekinumab Crohn's Disease Study Group. 2008. A randomized trial of ustekinumab, a human interleukin-12/23 monoclonal antibody, in patients with moderate-to-severe Crohn's disease. *Gastroenterology* **135:** 1130–1141.

46. Noonan, C.W., S.J. Kathman & M.C. White. 2002. Prevalence estimates for MS in the United States and evidence of an increasing trend for women. *Neurology* **58:** 136–138.

47. Segal, B.M. & E.M. Shevach. 1996. IL-12 unmasks latent autoimmune disease in resistant mice. *J. Exp. Med.* **184:** 771–775.

48. Langrish, C.L., Y. Chen, W.M. Blumenschein, *et al.* 2005. IL-23 drives a pathogenic T cell population that induces autoimmune inflammation. *J. Exp. Med.* **201:** 233–240.

49. Leonard, J.P., K.E. Waldburger & S.J. Goldman. 1995. Prevention of experimental autoimmune encephalomyelitis by antibodies against interleukin 12. *J. Exp. Med.* **181:** 381–386.

50. Constantinescu, C.S., M. Wysocka, B. Hilliard, *et al.* 1998. Antibodies against IL-12 prevent superantigen-induced and spontaneous relapses of experimental autoimmune encephalomyelitis. *J. Immunol.* **161:** 5097–5104.

51. Brok, H.P.M., M. van Meurs, E. Blezer, *et al.* 2002. Prevention of experimental autoimmune encephalomyelitis in common marmosets using an anti-IL-12p40 monoclonal antibody. *J. Immunol.* **169:** 6554–6563.

52. Cua, D.J., J. Sherlock, Y. Chen, *et al.* 2003. Interleukin-23 rather than interleukin-12 is the critical cytokine for autoimmune inflammation of the brain. *Nature* **421:** 744–748.

53. Windhagen, A., J. Newcombe, F. Dangond, *et al.* 1995. Expression of costimulatory molecules B7-1 (CD80), B7-2 (CD86), and interleukin 12 cytokine in multiple sclerosis lesions. *J. Exp. Med.* **182:** 1985–1996.

54. Li, Y., N. Chu, A. Hu, *et al.* 2007. Increased IL-23p19 expression in multiple sclerosis lesions and its induction in microglia. *Brain* **130:** 490–501.

55. Kouwenhoven, M., N. Teleshova., V. Özenci, *et al.* 2001. Monocytes in multiple sclerosis: phenotype and cytokine profile. *J. Neuroimmunol.* **112:** 197–205.

56. Kasper, L.H., D. Everitt, T.P. Leist, *et al.* 2006. A phase I trial of an interleukin-12/23 monoclonal antibody in relapsing multiple sclerosis. *Curr. Med. Res. Opin.* **22:** 1671–1678.

57. Segal, B.M., C.S. Constantinescu, A. Raychaudhuri, *et al.* 2008. Repeated subcutaneous injections of IL12/23 p40 neutralising antibody, ustekinumab, in patients with relapsing-remitting multiple sclerosis: a phase II, double-blind, placebo-controlled, randomised, dose-ranging study. *Lancet Neurol.* **7:** 796–804.

58. Hoeve, M.A., N.D.L. Savage, T. de Boer, *et al.* 2006. Divergent effects of IL-12 and IL-23 on the production of IL-17 by human T cells. *Eur. J. Immunol.* **36:** 661–670.

59. Longbrake, E.E. & M.K. Racke. 2009. Why did IL-12/IL-23 antibody therapy fail in multiple sclerosis? *Expert Rev. Neurother.* **9:** 319–321.

60. Trinchieri, G. 2003. Interleukin-12 and the regulation of innate resistance and adaptive immunity. *Nat. Rev. Immunol.* **3:** 133–146.

61. Filipe-Santos, O., J. Bustamante, A. Chapgier, *et al.* 2006. Inborn errors of IL-12/23- and IFN-γ-mediated immunity: molecular, cellular, and clinical features. *Sem. Immunol.* **18:** 347–361.

62. Novelli, F. & J.L. Casanova. 2004. The role of IL-12, IL-23 and IFN-γ in immunity to viruses. *Cytokine Growth Factor Rev.* **15:** 367–377.

63. Guia, S., C. Cognet, L. de Beaucoudrey, *et al.* 2008. A role for interleukin-12/23 in the maturation of human natural killer and CD56[+] T cells in vivo. *Blood* **111:** 5008–5016.

# Blocking Interleukin-1 in Rheumatic Diseases

## Its Initial Disappointments and Recent Successes in the Treatment of Autoinflammatory Diseases

### Raphaela Goldbach-Mansky

*National Institute of Arthritis and Musculoskeletal and Skin Diseases at the National Institutes of Health, Bethesda, Maryland, USA*

The role of the potent proinflammatory cytokine IL-1 in disease could clinically be investigated with the development of the IL-1 blocking agent anakinra (Kineret®), a recombinant IL-1 receptor antagonist. It was first tested in patients with sepsis without much benefit but was later FDA approved for the treatment of patients with rheumatoid arthritis. More recently IL-1 blocking therapies are used successfully to treat a new group of immune-mediated inflammatory conditions, autoinflammatory diseases. These conditions include rare hereditary fever syndromes and pediatric and adult conditions of Still's disease. Recently the FDA approved two additional longer acting IL-1 blocking agents, for the treatment of cryopyrin-associated periodic syndromes (CAPS), an IL-1 dependent autoinflammatory syndrome. The study of autoinflammatory diseases revealed mechanisms of IL-1 mediated organ damage and provided concepts to a better understanding of the pathogenesis of more common diseases such as gout and Type 2 diabetes which show initial promising results with IL-1 blocking therapy.

*Key words:* interleukin (IL-) 1; autoinflammatory diseases; neonatal onset multisystem inflammatory disease (NOMID); chronic inflammatory neurologic, cutaneous and arthritis syndrome (CINCA); deficiency of the IL-1 receptor antagonist (DIRA); *NLRP3*; *IL1RN*; IL-1Ra; anakinra; neonatal disorder; genetic disease

## History and Background

Interleukin1 (IL-1) is the prototype of a proinflammatory "alarm" cytokine that coordinates responses to endogenous and exogenous danger to the organism; it particularly coordinates the immune and hematologic responses. IL-1 was the first member of the family of IL-1 receptor molecules which currently consists of 11 members. IL-1α and IL-1β, both bind the biologically active IL-1 receptor (IL-1R) Type I and the inactive receptor Type II. To form an active signaling complex, the IL-1R Type I bound to either IL-1β or α, must associate with the accessory protein (IL-1RAcP) (Fig. 1). In 1986 a soluble factor was isolated from the urine of female patients that blocked the binding of IL-1 to its receptor.[1] Four years later this factor, the IL-1 receptor antagonist (IL-1Ra), was purified and cloned as the first naturally occurring receptor antagonist that blocked the action of a cytokine.[2] IL-1Ra is also a member of the IL-1 family and has 26 to 30% homology with the gene structure of IL-1β and 19% with that of IL-1α. Similar to the gene location of IL-1α und β, the gene location of *IL1RN* is also on the human chromosome 2q14.[3] As demonstrated in (Fig. 1), IL-1Ra inhibits the formation of an IL-1 signaling complex and is an important negative regulator. The balance between IL-1 and IL-1Ra is strictly regulated at many levels, and an imbalance of IL-1 and IL-1Ra has been implicated as the cause or a severity factor in a number of diseases.[4] The discovery of two different autoinflammatory diseases that are mediated

Address for correspondence: Raphaela Goldbach-Mansky, M.D., M.H.S., NIH/NIAMS, Building 10, Rm. 6D-47B, 10 Center Dr., Bethesda, MD 20892. goldbacr@mail.nih.gov

Cytokine Therapies: Ann. N.Y. Acad. Sci. 1182: 111–123 (2009).
doi: 10.1111/j.1749-6632.2009.05159.x © 2009 New York Academy of Sciences.

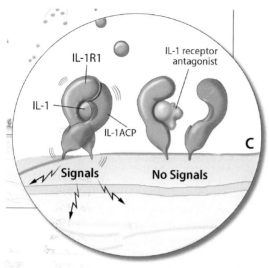

**Figure 1.** IL-1 receptor signaling. IL-1α and IL-1β can bind to the IL-1R1 receptor which recruits the accessory receptor. This receptor complex forms a signaling unit (ACP). However binding of the IL-1 receptor antagonist to the IL-1R1 receptor inhibits IL-1 binding and does not allow for association with the ACP and therefore no signaling through the receptor occurs.

by two distinct genetic abnormalities in the IL-1 pathway has demonstrated the phenotypic manifestations of a dysregulated IL-1 pathway in human disease.

The majority of IL-1α is bound to the plasma membrane on monocytes and B cells or remains inside the cell and may serve as an autocrine growth factor and as a DNA-binding transcription factor while IL-1β is produced primarily by macrophages upon stimulation with microbial and nonmicrobial factors. Both cytokines lack a leader sequence. The events necessary to lead to the secretion of IL-1α and IL-1β are still incompletely understood. IL-1β is activated via an "inflammasome complex," a molecular cytoplasmic platform that activates pro-caspase-1, the enzyme that cleaves pro- IL-1β in to its active form.[5] IL-1α and IL-1β differ in their patterns of organ distribution. IL-1α is expressed in high levels (higher than IL-1β) in lymphoreticular organs, intestine, spleen liver and lungs; it is dominantly detected in the lumen facing epithelial cells, while IL-1β is ex-

pressed in higher levels than IL-1α in privileged organs, such as kidney, heart, skeletal muscle and brain.[6] IL-1β gets rapidly activated and may coordinate a more restricted inflammatory response, it leads to induction of IL-1β itself, TNF-α, MMPs, iNOS, COX-2 and PLA 2, depending on the target cell type. IL-1β is the most powerful endogenous pyrogen known and has been implicated in tissue damage and systemic symptoms of sepsis and several inflammatory diseases.[4] Therefore IL-1 has been an early target in the therapy of a number of diseases but its path through the recent medical history has been marked by failures and successes.

## Clinical Studies with the IL-1 Receptor Antagonist

### Sepsis

Early studies in the late 1980s suggested that IL-1 levels were elevated in patients with sepsis and levels correlated with mortality. Blockade of IL-1β has attenuated the severity of disease and mortality in experimental models of shock and sepsis.[7] Although blockade of IL-1 was effective in animal models, two major studies in sepsis have provided conflicting results concerning the morbidity and mortality in humans. A phase II study in patients with sepsis suggested that treatment with the recombinant IL-1 RA, anakinra, reduced 28-day all cause- mortality in a dose-dependent manner, however a phase III trial failed to demonstrate a reduction in the 28-day mortality (Table 1). Despite the negative effect on overall mortality, subset analysis suggests that patients in shock or with gram negative sepsis are likely to most benefit from treatment. These results await further clarification and confirmation.

### Rheumatoid Arthritis

Rheumatoid arthritis (RA), a chronic autoimmune disease is characterized by chronic

**TABLE 1.** "Selected" Clinical Studies in Sepsis and Rheumatoid Arthritis

| Reference | Number of patients | Treatment intervention | Outcome | Safety | Other |
|---|---|---|---|---|---|
| **Sepsis Trials** | | | | | |
| Fisher et al.[21] | $n = 99$ recruited by 12 US centers | open label, PLB or 100 mg loading and 72 h infusion of: IL-1ra 17, 67 or 133 mg/h | Evaluation of mortality at day 28: PLB: (44%); 17 mg/hr (32%); 67 mg/hr (25%); 133 mg/hr (16%) | well tolerated | benefit in patients with high IL-6 baseline levels |
| Fisher et al.[22] | $n = 893$ | RCT, PLB or ANAK 100 mg loading and 72 hr infusion of: IL-1ra (1.0 or 2.0 mg/kg/h) | Evaluation of mortality at day 28: No significant difference between PLB and treatment arms | well tolerated | survival benefit in retrospective analysis of patients with one or more organ dysfunction |
| **Rheumatoid Arthritis Trials** | | | | | |
| Bresnihan et al.[23]; Jiang et al.[24] | $n = 472$ | RCT, PLB or ANAK at (30, 75 or 150 mg/day) | Evaluation of ACR 20 at 24 wks: 27% vs. 33%, 34%, 43% | well tolerated | local injection site reactions are common |
| Cohen et al.[25] | $n = 419$ | RCT, MTX + PLB or MTX + ANAK (0.04, 0.1, 0.4, 1 or 2 mg/kg/day) | Evaluation of ACR 20 at 12 wks: 23% vs. 19%, 30%, 36%, 42%, 35% | well tolerated | local injection site reactions are common |
| Cohen et al.[26] | $n = 501$ | RCT, MTX + PLB or MTX + ANAK (100 mg/day) | ACR 20: 22% vs. 38%; ACR 50: 8% vs. 17%; ACR 70: 2% vs. 6% | well tolerated | local injection site reactions are common |
| Fleischmann et al.[27,28] | $n = 1399$ | Phase IV, DMARD + PLB or DMARD + ANAK (100 mg/day) | | well tolerated | local injection site reactions are common |

PLB = placebo; RCT = randomized controlled trial; ANAK = anakinra; ACR = American College of Rheumatology criteria of disease improvement in rheumatoid arthritis; MTX = methotrexate; DMARD = Disease modifying antirheumatic drug.

inflammation of the joint lining (synovial membrane) which causes pain and swelling of multiple joints, primarily of the small joints of hands, feet and wrists. Over time, uncontrolled disease results in progressive joint damage, disability and increased mortality. Patients can be effectively treated with disease modifying antirheumatic drugs (DMARD) including methotrexate and leflunomide. The evolving understanding of the immune mechanisms that perpetuate the inflammatory response has led to the development of effective targeted therapies that have revolutionized the treatment of patients with active RA. Approved biologics for the treatment of RA include TNF-α blockade, elimination of B cells and blocking of co-stimulatory pathways.[8] IL-1 blockade has also been evaluated in the treatment of RA and the clinical trials are summarized in Table 1. Treatment with anakinra is well tolerated, opportunistic infections are rare compared to those seen with anti-TNF agents, and injection site reactions are the most common side effects. Treatment with anakinra is more effective than placebo, and addition of anakinra to methotrexate therapy in patients with an inadequate response to methotrexate alone, significantly improved joint swelling, pain and inflammatory blood markers as measured by ACR 20, ACR 50, and ACR 70 responses. However the inconvenience of daily injections of anakinra and the superior clinical effect of other anticytokine therapies have made anakinra a less favored choice in the treatment of rheumatoid arthritis.[9] The clinical data of IL-1 blockade with anakinra in RA have raised the questions whether anakinra is not a "good enough drug" to block IL-1 in RA or whether targeting IL-1 in RA is the wrong cytokine.

## Autoinflammatory Diseases

### Monogenic Autoinflammatory Diseases

The discovery of the genetic causes of a rare group of diseases that were initially termed hereditary fever syndromes, led to the definition of a new group of diseases which are now called "autoinflammatory diseases." Autoinflammatory diseases are a new group of immune dysregulatory disorders that are distinct from infections, allergic diseases, immunodeficiencies, and autoimmune diseases and are characterized clinically by recurrent episodes of systemic inflammation (elevation of acute phase reactants), predominance of neutrophils in the inflammatory infiltrate, and by organ-specific inflammation that can affect the skin, joints, bones, eyes, gastrointestinal tract, inner ears, and the central nervous system. Infection, autoantibodies and antigen-specific T cells are not identified in patients. Single gene mutations in a subset of the "monogenic" autoinflammatory disorders have been identified in familial Mediterranean fever, PAPA (pyogenic arthritis, pyoderma gangrenosum, and acne) syndrome and the cryopyrin-associated periodic syndromes (CAPS), pediatric granulomatous arthritis (PGA), and others. The identification of these genes has helped to pinpoint dysregulated inflammatory pathways in the in-

nate immune system and have linked the pathogenesis of these disorders to exaggerated responses to endogenous or exogenous "danger" triggers.[10] The currently identified autoinflammatory diseases with known genetic causes are listed in Table 2.

### CAPS and DIRA

Two disorders that are caused by dysregulated IL-1 responses with remarkable clinical responses to IL-1 blockade will be discussed in detail in this paper, these include the spectrum of cryopyrin associated periodic syndromes (CAPS) that is caused by mutations in *NLRP3, NALP3*[11–13] and deficiency of the IL-1 receptor antagonist (DIRA) that is caused by homozygous mutations in the IL-1 receptor antagonist gene (*IL1RN*).[14,15] *NLRP3* (also *NALP3*) encodes a protein, cryopyrin, that is a major component of the NLRP3 (NALP3 or cryopyrin) inflammasome, a macromolecular complex that activates caspase-1 the enzyme that controls activation and secretion of bioactive IL-1$\beta$ (Fig. 2). The inflammasome can be triggered by a number of exogenous stimuli or "danger signals" that include conserved microbial components and large inorganic crystalline structures such as asbestos and silica, but also endogenous "danger signals" that get released for example when cells are stressed or are dying and include uric acid.

Laboratory research and clinical investigations conducted in parallel revealed the pivotal role of IL-1$\beta$ in causing the clinical disease phenotype of CAPS. Historically CAPS is described as three diseases, familial cold autoinflammatory syndrome (FCAS), Muckle Wells syndrome (MWS), and neonatal onset multisystem inflammatory disease (NOMID) also called chronic inflammatory neurologic, cutaneous and arthritis syndrome (CINCA). The discovery that genetic mutations in exon 3 of the gene, *NLRP3* or *NALP3* cause all three disease phenotypes have revealed that these syndromes are caused by IL-1$\beta$ overproduction and form a disease spectrum with FCAS being the mildest and NOMID the most severe disease manifestation. The disease is autosomal dominantly inherited and a history of other affected family members can be obtained from most patients with FCAS and MWS whereas NOMID/CINCA is caused by sporadic mutations and no family history of CAPS is present.

All CAPS patients present with episodes of fever, urticarial rash, joint pain, and elevations in acute phase reactants but differ in the spectrum of multiorgan disease manifestations and in long-term morbidity and mortality. In FCAS the inflammatory episodes are triggered by cold, can present outside of the neonatal period and flares last for 12–24 h long-term outcome is favorable and amyloidosis is rare. In MWS and NOMID/CINCA, episodes of fever, urticarial rash, and arthritis are continuous and not provoked by cold and disease is usually present at or around birth. Conjunctivitis, episcleritis, anterior ureitis and optic disc edema are also seen; and in a European cohort, amyloidosis was reported in up to 25% of patients with MWS. In MWS progressive neurosensory hearing loss presents in the 2nd to 3rd decade. In NOMID up to 60% of patients present with abnormal bony overgrowth and all patients have significant CNS inflammation. Physical disability in patients with NOMID is caused by joint contractures and severe growth retardation. Cognitive impairment is secondary to perinatal complications and central nervous system (CNS) inflammation, which includes chronic aseptic meningitis, the development of ventriculomegaly, cerebral atrophy, and seizures. Sensorineural hearing loss develops in most patients in the first decade of life, and progressive vision loss can be a consequence of optic nerve atrophy caused by chronically increased intracranial pressures. Other findings include short stature, frontal bossing, and rarely, flattening of the nasal bridge. If untreated, the reported mortality is estimated to be around 20% before patients reach adulthood.[16]

We recently found that homozygous mutations of *IL1RN*, the gene encoding the IL-1 receptor antagonist (IL-1Ra), cause a severe inflammatory disease with some similarity to NOMID which has a high mortality in childhood. Mutations in *IL1RN* lead to complete absence of IL-1Ra and thus unopposed action of IL-1 on the IL-1 receptor and presents with systemic inflammation, skin pustulosis and mulitfocal osteomyelitis. Vasculitis and pulmonary manifestations can occur. Patients with DIRA do not have CNS or inner ear inflammation and respond dramatically to treatment with anakinra which is the very protein these children are missing. The *IL1RN* mutations are present in founder populations in Newfoundland, the Netherlands, and Puerto Rico and possibly Lebanon,[14,15] and further founder mutations have since been identified in two other populations (personal communications). Heterozygous carriers are asymptomatic and have no detectable cytokine abnormalities *in vitro*. Interestingly DIRA expands the spectrum of organs that can be damaged by increased IL-1 signaling to bone inflammation and pustular skin lesions (Fig. 3).

## Clinical Response to Treatment with IL-1 Blocking Agents in NOMID and DIRA

The clinical results to IL-1 blockade are striking; patients with CAPS respond well to treatment with anakinra and more recently the newer long acting IL-1 inhibitors (Table 3).

Clinical studies have shown significant improvement in the clinical symptoms of CAPS, including rash, headaches, fevers, and joint pain and marked improvement in inflammatory markers with remission in many patients, remission is also seen in 60% of patients with NOMID. IL-1 blockade with anakinra in NOMID can reverse organ inflammation imaged on MRI including CNS leptomeningitis and cochlear inflammation which is the cause for progressive hearing loss.[17] Preliminary data in very young children suggest that disability may be prevented if therapy can be initiated early in life which requires early diagnosis (our own unpublished data). The dose of anakinra needed to suppress inflammation in CAPS depends on disease severity and clinical phenotype and is

**TABLE 2.** "Monogenic" Autoinflammatory Syndromes

| Disease | Clinical description/ Year mutation published | Gene | Protein | Inheritance pattern | Disease onset | Flare/fever pattern | Specific organ inflammation | Treatment |
|---|---|---|---|---|---|---|---|---|
| FMF (MIM 249100) | 1945/1997 | *MEFV* (16p13) | pyrin | autosomal recessive | 80% of the cases occur before the age of 20 | 1–3 days | skin, joints, peritoneum, pleura | colchicine, rarely IL-1 and TNF blockade or thalidomide if colchicine resistant |
| TRAPS (MIM 191190) | 1982/1999 | *TNFRSF1A* (12p13) | TNF receptor 1 | autosomal dominant | median age at onset 3 yrs | 1–4 wk | skin, eyes, joints, peritoneum, pleura | TNF blockade, steroids, IL-1 blockade, colchicine is ineffective |
| CAPS: | | | | | | | | |
| FCAS (MIM 120100) | 1940/2001 | *CIAS1* (1q44) | Cryopyrin | autosomal dominant | first 6 months of life, cold induced | <24 h | skin, eyes, joints | IL-1 blockade |
| MWS (MIM 191900) | 1961/2001 | *CIAS1* (1q44) | Cryopyrin | autosomal dominant | Infancy to adolescence | 24–48 h | skin, eyes, joints, inner ears, meninges (mild) | IL-1 blockade |
| NOMID (MIM 607115) | 1975/2002 | *CIAS1* (1q44) | Cryopyrin | autosomal dominant/ de novo | neonatal or early infancy | continuous with flares | skin, eyes, joints, inner ears, meninges, bony epiphyseal hyperplasia | IL-1 blockade |
| HIDS (MIM 260920) | 1984/1999 and 2000* | *MVK* (12q24) | Mevalonate kinase | autosomal recessive | median age at onset 6 months | 3–7 days | skin, eyes, joints, prominent lymph nodes | NSAIDS, corticosteroids, TNF and IL-1 blockade |

*Continued*

**TABLE 2.** *Continued*

| Disease | Clinical description/ Year mutation published | Gene | Protein | Inheritance pattern | Disease onset | Flare/fever pattern | Specific organ inflammation | Treatment |
|---|---|---|---|---|---|---|---|---|
| PGA (MIM 186580) | 1985/2001 and 2005** | *NOD2* (16q12) | Nod2 | autosomal dominant/ de novo | early childhood | uncommon | skin, eyes, joints | NSAIDS, Corticosteroids, methotrexate, cyclosporine, TNF or IL-1blockade |
| PAPA (MIM 604416) | 1997/2002 | *CD2BP1* (15q24) | PSTPIP1 | autosomal dominant | early childhood | common | skin, joints | Local and systemic corticosteroids, TNF or IL-1 blockade |
| Majeed's syndrome (MIM 609628) | 1989/2005 | *LPIN2* (18p11) | Lipin2 | autosomal recessive | early infancy (1–19 months) | weeks-months | bones, periosteum, anemia | NSAIDS, corticosteroids, IFNα |
| Cherubism (MIM 118400) | 1965/2001 | *SH3BP2* (4p16) | SH3BP2 | autosomal dominant | childhood, spontaneous remission by 3rd decade | uncommon | jaws, eyes (rare) | NSAIDS, TNF inhibition, interferon-α, azithromycin, bisphosphonates |
| FCAS2 (MIM 611762) | 2008/2008 | *NLRP12* (19q13) | NLRP12 (NALP12) | autosomal dominant | childhood, cold induced | 2–10 day; 1–3x per mo | skin, hearing, joints, Aphtous ulcers | corticosteroids, IL-1 blockade not tested |
| DIRA (MIM 612852) | 1985/2009 | *IL1RN* (2q14) | IL-1 receptor antagonist | autosomal recessive | neonatal or early infancy | continuous with flares | skin, bones, lungs (rare), vasculitis (rare) | Anakinra |

FMF—familial Mediterranean fever; TRAPS—tumor necrosis factor receptor–associated periodic syndromes; FCAS—familial cold autoinflammatory syndrome; MWS—Muckle-Wells syndrome; NOMID—neonatal onset multisystem inflammatory disease; CAPS—cryopyrin-associated periodic syndromes; HIDS—hyperimmunoglobulin D syndrome; PGA—pediatric granulomatous arthritis encompasses the familial Blau syndrome (MIM 186580) and the sporadic early onset sarcoidosis (MIM 609464); PAPA—pyogenic arthritis, pyoderma gangrenosum, and acne syndrome; DIRA–deficiency of the IL-1-receptor-antagonist, caused by autosomal-recessive "loss of function" mutations of *IL1RN*; MIM—Mendelian inheritance in man number.

*Two groups identified the gene in 1999 and 2000.

**The gene for the familial disease, Blau syndrome, was identified in 2001 and for the sporadic form, sporadic early onset sarcoidosis in 2005.

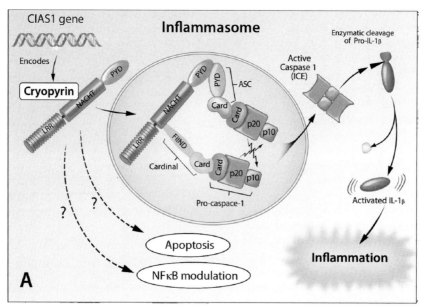

**Figure 2.** The Inflammasome, an IL-1 activating platform. Cryopyrin (NLRP3, NALP3, CIAS1) is a key molecule in regulating an inflammatory cytokine processing platform. Cryopyrin, ASC, Cardinal and two procaspase-1 molecules assemble to form, the cryopyrin inflammasome that activates caspase-1. Active caspase-1, enzymatically cleaves inactive IL-1β into its active form.

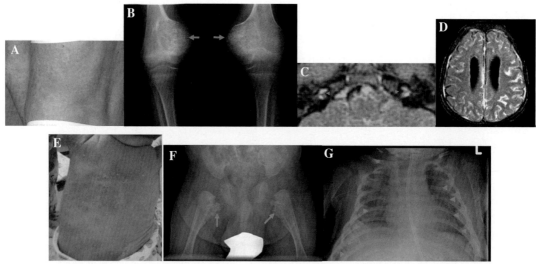

**Figure 3.** Clinical manifestations of NOMID/CINCA and DIRA. (**A**) to (**D**) depict characteristic clinical manifestations of patients with NOMID; (**E**) to (**G**) depict characteristic clinical manifestations of DIRA. (**A**) NOMID presents with an urticaria like rash, however the cellular infiltrate is neutrophilic consistent with neutrophilic dermatitis. (**B**) Radiographic findings of the knee show tumor like hyperostotic lesions originating in the growth plate. Once ossification is completed, the bone of these lesions is histologically normal. (**C**) Postcontrast FLAIR MRI of the inner ears shows abnormal cochlear enhancement suggestive of cochlear inflammation. (**D**) Postcontrast FLAIR MRI of brain show leptomeningeal enhancement. (**E**) shows generalized pustulosis seen in a 3-month-old infant. (**F**) A hip X-ray shows heterotrophic ossification or periosteal cloaking of the proximal femoral metaphysic and periosteal elevation of the diaphysis. (**G**) Typical radiographic manifestations on a chest X-ray include widening of multiple anterior ribs (arrows).

lowest in FCAS (0.5 to 1.5 mg/kg/day in most patients) up to 3.5 to 6 mg/kg/day in patients with NOMID/CINCA. Despite multiple open label studies showing the remarkable benefit of anakinra in CAPS, this drug has not been FDA approved for the treatment of these conditions. However recent successful drug development programs with the long acting IL-1 inhibitor Rilonacept, Arcalyst®, led to the first FDA approved therapy for CAPS.[18,19] A second long acting IL-1 inhibitor, canakinumab, Ilaris®, also showed efficacy in CAPS and was recently approved by the FDA for the treatment of CAPS.[20] Both agents were evaluated in patients with FCAS and MWS.

## IL-1 Blockade in Other Autoinflammatory Diseases

Anakinra has also been used to prevent attacks and reduce systemic inflammation in patients with colchicine resistant FMF, HIDS, and TRAPS, and clinical responses were also reported in patients with Blau and also PAPA syndrome.

In addition to the effect in monogenic diseases, a number of presumed polygenic autoinflammatory diseases have also successfully been treated with IL-1 inhibition (Table 3). Some likely polygenic autoinflammatory diseases that share clinical similarities with some monogenic autoinflammatory diseases show impressive responses to IL-1 blockade. These include acute and chronic gout, pseudogout and the management of Schnitzler syndrome, a rare acquired urticarial disease with clinical similarities to MWS that is also associated with a monoclonal IgM gammopathy. A subset of patients with pediatric and adult Morbus Still's disease is also responsive to IL-1 blockade. A case report of a patient with Behcet's disease and the improvement in glucose tolerance in Type II diabetes treated with IL-1 blockade suggest that IL-1 mediated organ damage is not only limited to a small subset of rare diseases (Table 3).

**TABLE 3.** Selected Clinical Studies in some Autoinflammatory Diseases

| Syndrome | Study |
| --- | --- |
| Monogenic Disorders* | |
| Familial Mediterranean Fever (FMF) | 29–31 |
| TNF receptor associated periodic syndrome (TRAPS) | 32–34 |
| Hyper IgD syndrome (HIDS) | 35,36 |
| CAPS (FCAS) | 37 |
| CAPS (Muckle Wells Syndrome) | 18–20 |
| CAPS (NOMID/CINCA) | 17 |
| Pediatric granulomatous arthritis (PCA)§ | 38 |
| PAPA syndrome¶ | 39 |
| Polygenic disorders** | |
| SOJIA*** | 40–45 |
| AOSD**** | 46–48 |
| Gout | 49,50 |
| Behcet's disease | 51 |
| Diabetes Type II | 52 |

*Monogenic disorders are caused by a homozygous or heterozygous mutations in genes associated with the modulating innate immune pathways.

§Pediatric granulomatous arthritis (PGA) is the term applied to the syndromes formerly described as Blau syndrome, a familial form of granulomatous disease, and early onset sarcoidosis, a sporadic form.

¶Pyogenic arthritis, pyoderma gangrenosum, and acne (PAPA) syndrome.

**These likely polygenetic diseases do not have any genetic mutations or polymorphisms identified yet but are believed to be caused by genetic predispositions.

***Only studies with more than 10 patients are listed; systemic onset juvenile idiopathic arthritis (SOJIA).

****Only studies with more than 5 patients are listed; adult onset still's disease (AOSD).

## Safety of IL-1 Therapy

The three currently approved drugs targeting IL-1 are listed in Table 4. All drugs are generally well tolerated. In two studies up to 71% of patients treated with anakinra developed an injection site reaction, which was typically reported within the first 4 weeks of therapy. The development of injection site reactions was uncommon after the first month of therapy. The incidence of infection was 40% in the anakinra-treated patients and 35% in placebo-treated patients. The incidence of serious infections in studies was 1.8% in anakinra-treated

**TABLE 4.** Currently Approved IL-1 Inhibitors

| Biologic | Class | Construct | Half-life | Onset of action | Binding target | Dose/ Administration |
|---|---|---|---|---|---|---|
| Anakinra* | IL-1 receptor antago-nist | recombinant human IL-1 receptor antago-nist | 4–6 h | 1–3 months in RA, within days in CAPS (pts with FCAS, MWS, NOMID/ CINCA) | IL-1 re-ceptor Type 1 | 100 mg sc daily not approved for use in children, used in children with CAPS: 1–5 mg/kg/day/sc |
| Rilonacept** | soluble IL-1 receptor-Ig | recombinant human IL-1 receptor –Ig fusion protein | 34–57 h | within days in CAPS (pts with FCAS and MWS) | IL-1α, IL-1β | loading dose of 320 mg then 160 mg sc weekly pediatric dose (for children older than 12 years): 4.4 mg/kg loading dose then 2.2 mg/kg/sc weekly |
| Canakinumab*** | anti IL-1β antibody | humanized anti IL-1β antibody | 26 days | within days in CAPS (pts with FCAS and MWS) | IL-1β | 150 mg sc every 8wks pediatric dose (for children older than 4 years): 2 mg/kg/sc every 8 wks |

*Anakinra (Kineret®) was approved by the U.S. FDA for the treatment of rheumatoid arthritis in 2001.

**Rilonacept (Arcalyst®) was approved by the U.S. FDA for the treatment of the orphan diseases CAPS in February 2008.

***Canakinumab (Ilaris®) was approved by the U.S. FDA for the treatment of the orphan diseases CAPS in July 2009.

patients and 0.6% in placebo-treated patients over six months. These infections consisted primarily of bacterial events such as celluli-tis, pneumonia, and bone and joint infections, rather than unusual, opportunistic, fungal, or viral infections. Most patients continued on study drug after the infection resolved. There were no on-study deaths due to serious infec-tious episodes in either study. In patients who received both anakinra and etanercept for up to 24 weeks, the incidence of serious infections was 7%. The most common infections consisted of bacterial pneumonia (4 cases) and cellulitis (4 cases). One patient with pulmonary fibro-sis and pneumonia died because of respiratory failure.[53]

The most commonly reported adverse reac-tion associated with Rilonacept was an injec-tion site reaction. In 360 patients treated with Rilonacept and 179 treated with placebo, the incidence of infections was 34% versus 27% for rilonacept and placebo. One Mycobacterium intracellulare infection after bursal injection and a death from Streptococcus pneumoniae meningitis occurred.[54] In the canakinumab studies injection site reactions occurred in up to 9% of patients and up to 14% of patients developed vertigo with the injections.[55]

## Summary

The discovery of the genetic causes for a number of monogenic autoinflammatory dis-eases in general and the discovery of mutations in *NLRP3/NALP3/CIAS1* that cause CAPS and mutations in *IL1RN* that cause DIRA in par-ticular, have resulted in profound advances in our understanding of the role of IL-1 in hu-man diseases. In CAPS the genetic defects lead to the oversecretion of IL-1β and in DIRA the absence of IL-1 receptor antagonist leads to un-opposed signaling of IL-1α and IL-1β through the IL-1 receptor. Through clinical studies with IL-1 blocking agents, the pivotal role of IL-1 in causing not only the systemic inflammation but also the organ specific disease has linked the discovery of the genetic cause, the understanding of the immunopathogenesis with the choice of a rational treatment approach. Following the disappointment of anti IL-1 therapy in sepsis and rheumatoid arthritis, the successes of anti-IL 1 therapy in the treatment of monogenic autoinflammatory diseases have not only clearly demonstrated the prominent role of IL-1 in human diseases, but also illustrate the success of molecular biology in exploring human diseases, and in developing a medical practice guided by our understanding of the disease pathogenesis and the rational use of targeted therapies. The expansion of the role of IL-1 in genetically complex diseases is justified by recent studies showing benefit of using IL-1 blockade in gout and Type II diabetes. Better equipped with the availability of novel long-acting biologics that block the IL-1 pathway, and the development of small orally administered molecules that target the IL-1 pathway and the inflammasome, the exploration of the IL-1 pathway in a broad spectrum of human diseases can continue.

## Conflicts of interest

The author declares no conflicts of interest.

## References

1. Balavoine, J.F., B. De Rochemonteix, K. Williamson, *et al*. 1986. Prostaglandin E2 and collagenase production by fibroblasts and synovial cells is regulated by urine-derived human interleukin 1 and inhibitor(s). *J. Clin. Invest.* **78:** 1120–1124.
2. Hannum, C.H., C.J. Wilcox, W.P. Arend, *et al*. 1990. Interleukin-1 receptor antagonist activity of a human interleukin-1 inhibitor. *Nature* **343:** 336–340.
3. Dinarello, C.A. 1996. Biologic basis for interleukin-1 in disease. *Blood* **87:** 2095–2147.
4. Arend, W.P. 2002. The balance between IL-1 and IL-1Ra in disease. *Cytokine Growth Factor Rev.* **13:** 323–340.
5. Petrilli, V., C. Dostert, D.A. Muruve & J. Tschopp. 2007. The inflammasome: a danger sensing complex triggering innate immunity. *Curr. Opin. Immunol.* **19:** 615–622.
6. Hacham, M., S. Argov, R.M. White, *et al*. 2000. Distinct patterns of IL-1 alpha and IL-1 beta organ distribution–a possible basis for organ mechanisms of innate immunity. *Adv. Exp. Med. Biol.* **479:** 185–202.
7. Cain, B.S., D.R. Meldrum, A.H. Harken & R.C. Mcintyre, Jr. 1998. The physiologic basis for anticytokine clinical trials in the treatment of sepsis. *J. Am. Coll. Surg.* **186:** 337–350.
8. Smolen, J.S., D. Aletaha, M. Koeller, *et al*. 2007. New therapies for treatment of rheumatoid arthritis. *Lancet* **370:** 1861–1874.
9. O'Dell, J.R. 2004. Therapeutic strategies for rheumatoid arthritis. *N. Engl. J. Med.* **350:** 2591–2602.
10. Glaser, R.L. & R. Goldbach-Mansky. 2008. The spectrum of monogenic autoinflammatory syndromes: understanding disease mechanisms and use of targeted therapies. *Curr. Allergy Asthma Rep.* **8:** 288–298.
11. Hoffman, H.M., J.L. Mueller, D.H. Broide, *et al*. 2001. Mutation of a new gene encoding a putative pyrin-like protein causes familial cold autoinflammatory syndrome and Muckle-Wells syndrome. *Nat. Genet.* **29:** 301–305.
12. Aksentijevich, I., M. Nowak, M. Mallah, *et al*. 2002. De novo CIAS1 mutations, cytokine activation, and evidence for genetic heterogeneity in patients with neonatal-onset multisystem inflammatory disease (NOMID): a new member of the expanding family of pyrin-associated autoinflammatory diseases. *Arthritis Rheum.* **46:** 3340–3348.
13. Feldmann, J., A.M. Prieur, P. Quartier, *et al*. 2002. Chronic infantile neurological cutaneous and articular syndrome is caused by mutations in CIAS1, a gene highly expressed in polymorphonuclear cells and chondrocytes. *Am. J. Hum. Genet.* **71:** 198–203.
14. Aksentijevich, I., S.L. Masters, P.J. Ferguson, *et al*. 2009. An autoinflammatory disease with deficiency of the interleukin-1-receptor antagonist. *N. Engl. J. Med.* **360:** 2426–2437.
15. Reddy, S., S. Jia, R. Geoffrey, *et al*. 2009. An autoinflammatory disease due to homozygous deletion of the IL1RN locus. *N. Engl. J. Med.* **360:** 2438–2444.
16. Prieur, A.M., C. Griscelli, F. Lampert, *et al*. 1987. A chronic, infantile, neurological, cutaneous and articular (CINCA) syndrome. A specific entity analysed in 30 patients. *Scand. J. Rheumatol. Suppl.* **66:** 57–68.
17. Goldbach-Mansky, R., N.J. Dailey, S.W. Canna, *et al*. 2006. Neonatal-onset multisystem inflammatory disease responsive to interleukin-1beta inhibition. *N. Engl. J. Med.* **355:** 581–592.

18. Goldbach-Mansky, R., S.D. Shroff, M. Wilson, *et al.* 2008. A pilot study to evaluate the safety and efficacy of the long-acting interleukin-1 inhibitor rilonacept (interleukin-1 Trap) in patients with familial cold autoinflammatory syndrome. *Arthritis Rheum.* **58:** 2432–2442.

19. Hoffman, H.M., M.L. Throne, N.J. Amar, *et al.* 2008. Efficacy and safety of rilonacept (interleukin-1 Trap) in patients with cryopyrin-associated periodic syndromes: results from two sequential placebo-controlled studies. *Arthritis Rheum.* **58:** 2443–2452.

20. Lachmann, H.J., I. Kone-Paut, J.B. Kuemmerle-Deschner, *et al.* 2009. Use of canakinumab in the cryopyrin-associated periodic syndrome. *N. Engl. J. Med.* **360:** 2416–2425.

21. Fisher, C.J., Jr., G.J. Slotman, S.M. Opal, *et al.* 1994. Initial evaluation of human recombinant interleukin-1 receptor antagonist in the treatment of sepsis syndrome: a randomized, open-label, placebo-controlled multicenter trial. *Crit. Care Med.* **22:** 12–21.

22. Fisher, C.J., Jr., J.F. Dhainaut, S.M. Opal, *et al.* 1994. Recombinant human interleukin 1 receptor antagonist in the treatment of patients with sepsis syndrome. Results from a randomized, double-blind, placebo-controlled trial. Phase III rhIL-1ra Sepsis Syndrome Study Group. *JAMA* **271:** 1836–1843.

23. Bresnihan, B., J.M. Alvaro-Gracia, M. Cobby, *et al.* 1998. Treatment of rheumatoid arthritis with recombinant human interleukin-1 receptor antagonist. *Arthritis Rheum.* **41:** 2196–2204.

24. Jiang, Y., H.K. Genant, I. Watt, *et al.* 2000. A multicenter, double-blind, dose-ranging, randomized, placebo-controlled study of recombinant human interleukin-1 receptor antagonist in patients with rheumatoid arthritis: radiologic progression and correlation of Genant and Larsen scores. *Arthritis Rheum.* **43:** 1001–1009.

25. Cohen, S., E. Hurd, J. Cush, *et al.* 2002. Treatment of rheumatoid arthritis with anakinra, a recombinant human interleukin-1 receptor antagonist, in combination with methotrexate: results of a twenty-four-week, multicenter, randomized, double-blind, placebo-controlled trial. *Arthritis Rheum.* **46:** 614–624.

26. Cohen, S.B., L.W. Moreland, J.J. Cush, *et al.* 2004. A multicentre, double blind, randomised, placebo controlled trial of anakinra (Kineret), a recombinant interleukin 1 receptor antagonist, in patients with rheumatoid arthritis treated with background methotrexate. *Ann. Rheum. Dis.* **63:** 1062–1068.

27. Fleischmann, R.M., J. Schechtman, R. Bennett, *et al.* 2003. Anakinra, a recombinant human interleukin-1 receptor antagonist (r-metHuIL-1ra), in patients with rheumatoid arthritis: A large, international, multi-center, placebo-controlled trial. *Arthritis Rheum.* **48:** 927–934.

28. Fleischmann, R.M., J. Tesser, M.H. Schiff, *et al.* 2006. Safety of extended treatment with anakinra in patients with rheumatoid arthritis. *Ann. Rheum. Dis.* **65:** 1006–1012.

29. Gattringer, R., H. Lagler, K.B. Gattringer, *et al.* 2007. Anakinra in two adolescent female patients suffering from colchicine-resistant familial Mediterranean fever: effective but risky. *Eur. J. Clin. Invest.* **37:** 912–914.

30. Kuijk, L.M., A.M. Govers, J. Frenkel & W.J. Hofhuis. 2007. Effective treatment of a colchicine-resistant familial Mediterranean fever patient with anakinra. *Ann. Rheum. Dis.* **66:** 1545–1546.

31. Roldan, R., A.M. Ruiz, M.D. Miranda & E. Collantes. 2008. Anakinra: new therapeutic approach in children with Familial Mediterranean Fever resistant to colchicine. *Joint Bone Spine* **75:** 504–505.

32. Simon, A., E.J. Bodar, J.C. Van Der Hilst, *et al.* 2004. Beneficial response to interleukin 1 receptor antagonist in traps. *Am. J. Med.* **117:** 208–210.

33. Gattorno, M., M.A. Pelagatti, A. Meini, *et al.* 2008. Persistent efficacy of anakinra in patients with tumor necrosis factor receptor-associated periodic syndrome. *Arthritis Rheum.* **58:** 1516–1520.

34. Sacre, K., B. Brihaye, O. Lidove, *et al.* 2008. Dramatic improvement following interleukin 1beta blockade in tumor necrosis factor receptor-1-associated syndrome (TRAPS) resistant to anti-TNF-alpha therapy. *J. Rheumatol.* **35:** 357–358.

35. Bodar, E.J., J.C. Van Der Hilst, J.P. Drenth, *et al.* 2005. Effect of etanercept and anakinra on inflammatory attacks in the hyper-IgD syndrome: introducing a vaccination provocation model. *Neth. J. Med.* **63:** 260–264.

36. Cailliez, M., F. Garaix, C. Rousset-Rouviere, *et al.* 2006. Anakinra is safe and effective in controlling hyperimmunoglobulinaemia D syndrome-associated febrile crisis. *J. Inherit. Metab. Dis.* **29:** 763.

37. Hoffman, H.M., S. Rosengren, D.L. Boyle, *et al.* 2004. Prevention of cold-associated acute inflammation in familial cold autoinflammatory syndrome by interleukin-1 receptor antagonist. *Lancet* **364:** 1779–1785.

38. Arostegui, J.I., C. Arnal, R. Merino, *et al.* 2007. NOD2 gene-associated pediatric granulomatous arthritis: clinical diversity, novel and recurrent mutations, and evidence of clinical improvement with interleukin-1 blockade in a Spanish cohort. *Arthritis Rheum.* **56:** 3805–3813.

39. Dierselhuis, M.P., J. Frenkel, N.M. Wulffraat & J.J. Boelens. 2005. Anakinra for flares of pyogenic arthritis in PAPA syndrome. *Rheumatology (Oxford)* **44:** 406–408.

40. Reiff, A. 2005. The use of anakinra in juvenile arthritis. *Curr. Rheumatol. Rep.* **7:** 434–440.

41. Pascual, V., F. Allantaz, E. Arce, *et al.* 2005. Role of interleukin-1 (IL-1) in the pathogenesis of systemic onset juvenile idiopathic arthritis and clinical response to IL-1 blockade. *J. Exp. Med.* **201:** 1479–1486.

42. Gattorno, M., A. Piccini, D. Lasiglie, *et al.* 2008. The pattern of response to anti-interleukin-1 treatment distinguishes two subsets of patients with systemic-onset juvenile idiopathic arthritis. *Arthritis Rheum.* **58:** 1505–1515.

43. Lequerre, T., P. Quartier, D. Rosellini, *et al.* 2008. Interleukin-1 receptor antagonist (anakinra) treatment in patients with systemic-onset juvenile idiopathic arthritis or adult onset Still disease: preliminary experience in France. *Ann. Rheum. Dis.* **67:** 302–308.

44. Quartier, P. 2008. [Still's disease (Systemic-Onset Juvenile Idiopathic Arthritis)]. *Arch. Pediatr.* **15:** 865–866.

45. Lovell, D.J., N. Ruperto, S. Goodman, *et al.* 2008. Adalimumab with or without methotrexate in juvenile rheumatoid arthritis. *N. Engl. J. Med.* **359:** 810–820.

46. Lequerre, T., P. Quartier, D. Rosellini, *et al.* 2008. Interleukin-1 receptor antagonist (anakinra) treatment in patients with systemic-onset juvenile idiopathic arthritis or adult onset Still disease: preliminary experience in France. *Ann. Rheum. Dis.* **67:** 302–308.

47. Fitzgerald, A.A., S.A. Leclercq, A. Yan, *et al.* 2005. Rapid responses to anakinra in patients with refractory adult-onset Still's disease. *Arthritis Rheum.* **52:** 1794–1803.

48. Vasques Godinho, F.M., M.J. Parreira Santos & D.S. Canas. 2005. Refractory adult onset Still's disease successfully treated with anakinra. *Ann. Rheum. Dis.* **64:** 647–648.

49. So, A., T. De Smedt, S. Revaz & J. Tschopp. 2007. A pilot study of IL-1 inhibition by anakinra in acute gout. *Arthritis Res. Ther.* **9:** R28.

50. McGonagle, D., A.L. Tan, S. Shankaranarayana, *et al.* 2007. Management of treatment resistant inflammation of acute on chronic tophaceous gout with anakinra. *Ann. Rheum. Dis.* **66:** 1683–1684.

51. Botsios, C., P. Sfriso, A. Furlan, *et al.* 2008. Resistant Behcet disease responsive to anakinra. *Ann. Intern. Med.* **149:** 284–286.

52. Larsen, C.M., M. Faulenbach, A. Vaag, *et al.* 2007. Interleukin-1-receptor antagonist in type 2 diabetes mellitus. *N. Engl. J. Med.* **356:** 1517–1526.

53. Anakinra package insert: http://www.kineretrx.com/professional/pi.jsp

54. Rilonacept package insert: http://www.regeneron.com/ARCALYST-fpi.pdf

55. Canakinumab package insert: http://www.pharma.us.novartis.com/products/name/ilaris.jsp

# Rilonacept—CAPS and Beyond

## A Scientific Journey

### Neil Stahl, Allen Radin, and Scott Mellis

*Regeneron Pharmaceuticals, Inc., Tarrytown, New York, USA*

Rilonacept is a dimeric fusion protein consisting of the extracellular domains of inter-leukin (IL)-1 type 1 receptor and IL-1 receptor accessory protein joined to the constant region (Fc) of human immunoglobulin G1. By incorporating both components of the IL-1 binding complex, rilonacept is able to tightly bind IL-1 with picomolar affinity. Although early clinical results in rheumatoid arthritis (RA) suggested that RA is not primarily an IL-1-driven disease, the discovery that the rare genetic conditions called cryopyrin-associated periodic syndromes (CAPS) were caused by overproduction of IL-1 led to clinical development and approval for these conditions. An assay that detects rilonacept:IL-1 complexes in plasma is helping to identify new indications, such as gout, in which IL-1 overproduction plays a key pathogenic role. The development of rilona-cept for CAPS was achieved through collaboration between the pharmaceutical industry, academia, and government agencies, and demonstrates that knowledge gleaned in or-phan indications can inform drug development for more common and heterogeneous diseases.

*Key words:* cryopyrin-associated periodic syndromes (CAPS); drug development; cy-tokines; rilonacept; interleukin-1; gout

## Introduction

Rilonacept (Arcalyst®) is a dimeric fusion protein, developed by Regeneron Pharmaceu-ticals and indicated for the treatment of a group of rare genetic diseases called cryopyrin-associated periodic syndromes (CAPS) caused by dysregulated interleukin (IL)-1 production. The story of rilonacept's development is one of a complex scientific journey, marked by careful bench research, serendipitous discovery, pur-suit of some therapeutic pathways that turned out to be "dead-ends" and collaboration be-tween the pharmaceutical industry, academia, and government agencies, such as the National Institutes of Health (NIH) and the Food and Drug Administration (FDA). All of these factors resulted in the rapid development and approval of the first agent to treat CAPS, conditions that were otherwise managed only by symptomatic therapies and avoidance of triggers, and which caused patients considerable discomfort and disruption to their daily lives. This review de-scribes the development of rilonacept and high-lights the way in which our evolving under-standing of cytokine biology is opening the door to many potential therapies for difficult-to-treat conditions.

## First Steps

Scientists at Regeneron began working on cytokine signaling in the 1990s. One of our key discoveries was that interaction of the cytokine with its membrane-bound receptor drives the formation of a multicomponent re-ceptor complex, and that formation of the complex is essential for activation of intra-cellular signaling.[1] Many cytokines bind with low affinity to the first receptor component ($\alpha$-component) and then recruit a second ($\beta$)

---

Address for correspondence: Neil Stahl, Regeneron Pharmaceuticals, 777 Old Saw Mill River Rd, Tarrytown, NY 10591. Voice: 914 345 7400; fax: 914 345 7650. neil.stahl@regeneron.com

Cytokine Therapies: Ann. N.Y. Acad. Sci. 1182: 124–134 (2009).
doi: 10.1111/j.1749-6632.2009.05074.x © 2009 New York Academy of Sciences.

**A**

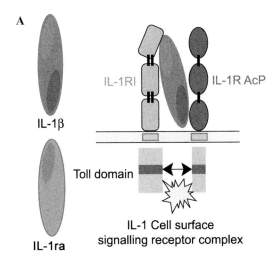

IL-1β

IL-1ra

IL-1RI        IL-1R AcP

Toll domain

IL-1 Cell surface
signalling receptor complex

**B**

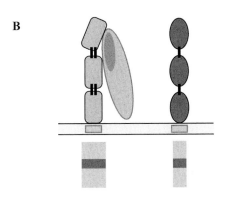

IL-1ra blocks dimerization
of cell surface signalling
receptor complex

**C**

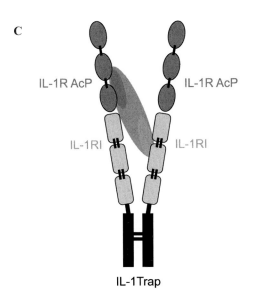

IL-1R AcP        IL-1R AcP

IL-1RI        IL-1RI

IL-1Trap

receptor component.[1,2] The cytokine is bound to the complex of two receptor components with much higher affinity than with either receptor component alone.[2] In the case of IL-1, binding occurs first to the type 1 receptor (IL-1R1) on the cell surface before recruitment of the IL-1 receptor accessory protein (IL-1RAcP) (Fig. 1A). We hypothesized that creating a soluble receptor containing both receptor components would bind IL-1 with high affinity, forming a potent antagonist. We created heterodimeric proteins by fusing each extracellular domain to the Fc portion of human immunoglobulin G1 (IgG1), which drives formation of disulfide-linked dimers. First-generation heterodimeric constructs were indeed able to bind IL-1 with high affinity, but were difficult to manufacture. This limitation was overcome by creation of an in-line fusion of the extracellular domains of IL-1RAcP to the mature extracellular domain of IL-1R1, which was then fused to the Fc domain.[2] This simple homodimeric construct incorporates both extracellular domains, which bind IL-1β and IL-1α with picomolar affinity, but is still easy to express and manufacture, and became rilonacept (Fig. 1C).[2,3]

During this period of research, the biologic therapies for rheumatoid arthritis (RA) were undergoing considerable clinical research.

**Figure 1.** The mechanism by which rilonacept blocks IL-1 activity.[3] (**A**) IL-1 has two receptor binding sites (blue and red). IL-1R1 binds to one site and IL-RAcP to the other. Cellular activation occurs when IL-1 binds to the complex containing both receptor components. (**B**) Naturally occurring IL-1R antagonist binds to only one site and prevents complex formation and therefore completes activation. (**C**) IL-1 Trap (rilonacept) contains the extracellular domains of both IL-1 receptor components fused to the Fc domains from human IgG1. Disulfide bonds between the Fc chains create a dimeric structure. IL-1 binds to rilonacept with high affinity, similar to the binding seen with the receptor complex on the cell surface. Adapted with permission from Dinarello, C.A. 2003. Setting the cytokine trap for autoimmunity. *Nat. Med.* **9:** 20–22. © Nature Publishing Group, 2003.

IL-1 was widely reported as an important cytokine in the pathogenesis of RA and the recombinant human IL-1 receptor antagonist (rhIL1ra) (Fig. 1B), anakinra, was under development for this indication. However, this molecule was known to have a relatively short half-life that necessitates once-daily parenteral administration. This prompted Regeneron to consider developing IL-1 Trap (now called rilonacept) for the treatment of RA, and early animal studies in RA models indicated a potent anti-inflammatory effect.[2] Anakinra was approved for the treatment of RA in 2001, however its efficacy in the treatment of RA was generally regarded to be lower than that of the TNF-α antagonists.[4] It was unknown whether this efficacy profile was due to potential pharmacological limitations of rhIL1ra or whether IL-1 was of less importance than other inflammatory cytokines as a therapeutic target in RA. Although it was hypothesized that rilonacept might be more clinically active than anakinra (based on high affinity and long duration of action), a Phase II study in RA indicated clinical efficacy that did not equal that of the TNF blockers,[5–8] and further clinical development of rilonacept for RA was suspended in 2004.

## A New Direction

At approximately the same time, in multiple locations, researchers were attempting to unravel the genetic determinants of a rare group of genetic syndromes now called "Cryopyrin Associated Periodic Syndromes (CAPS)".[9–11] CAPS include familial cold autoinflammatory syndrome (FCAS), Muckle-Wells syndrome (MWS), and neonatal-onset multisystem inflammatory disease (NOMID). The conditions share common features of spontaneous generalized painful or pruritic erythematous rash, fever, and flulike symptoms of headache, fatigue, myalgia, arthralgia, and leukocytosis consistent with systemic inflammation (Table 1). The intensity of disease and associated disability is progressive from FCAS to MWS to NOMID.[12] Patients with FCAS may be relatively asymptomatic upon awakening, however, exposure to modest degrees of cooling temperature change (e.g., a cool breeze or air conditioning) will induce a significant systemic inflammatory reaction within hours. Patients with MWS experience inflammation more constitutively and a significant number (60%) develop sensorineural hearing loss due

**TABLE 1.** Clinical Features of CAPS[12]

| | FCAS (least severe) | MWS (Intermediate) | NOMID (most severe) |
|---|---|---|---|
| Inheritance pattern | Autosomal dominant | Autosomal dominant | Autosomal dominant and sporadic |
| Age at onset | Infancy | Infancy and childhood | Infancy |
| Triggers | Generalized cold exposure | Spontaneous (no triggers), also cold, stress, exercise | None |
| Length of episodes | 1–2 days | 1–2 days with chronic symptoms | Chronic symptoms with flares |
| Skin features | Generalized urticaria-like rash | Generalized urticaria-like rash | Generalized urticaria-like rash |
| Joint involvement | Polyarthralgia | Polyarthralgia and polyarthritis | Polyarthritis and distal femur enlargement |
| Sensory features | Headache and conjunctivitis | Hearing loss and conjunctivitis | Hearing loss, papilledema, seizures, aseptic meningitis |
| Other features | Myalgia | Myalgia, amyloidosis | Organomegaly |

Adapted with kind permission of Current Medicine Group LLC from Hoffman, H.M. 2007. Hereditary immunologic disorders caused by pyrin and cryopyrin. *Curr. Allergy Asthma Rep.* **7:** 323–30. ©Current Medicine Group LLC, 2007.

to inflammation in the inner ear.[13] A significant number (25%) also develop renal amyloidosis, which is the primary cause of death in MWS adults.[12–14] NOMID has severe onset in early infancy, and patients may have central nervous system inflammation causing seizures, developmental delay and visual impairment, as well as characteristic deformity/overgrowth of the distal femur, proximal tibia and patella.[12] For patients with FCAS and MWS, symptoms wax and wane, with flares triggered by exposure to cold in the case of FCAS and by other more random triggers, potentially including cold, exercise or stress, in MWS. NOMID patients have generalized symptoms, which worsen with no apparent trigger.[12] The conditions are rare, affecting fewer than one in a million people,[15] but tend to run in families along a pattern of autosomal dominant inheritance.[12]

Using positional cloning, it was discovered that FCAS and MWS patients shared a number of mutations in a gene currently termed *NLRP3* (Nucleotide-binding domain, Leucine-Rich family, Pyrin domain containing 3) gene (previously known as *CIAS1*) on chromosome 1q44.[10,11,16] This gene codes for the cytoplasmic protein cryopyrin, which forms a complex with intracellular adaptor proteins, including ASC (apoptosis-associated Speck–like protein containing caspase activation and recruitment domains), CARDINAL (caspase activation and recruitment domain inhibitor of NF-κB-activating ligands), and caspase-1 (Fig. 2).[17] Together, this protein complex (found in macrophages and monocytes) is known as the inflammasome and is thought to be inactive in its native state.[18] When activated by pathogenic RNA, extracellular toxins, or other endogenous signals, the inflammasome converts inactive cytosolic pro-interleukin (IL)-1β into active IL-1β for secretion (Fig. 2).[19,20] Shortly after the discovery of the genetics of FCAS and MWS, a team at the National Institute of Arthritis and Musculoskeletal and Skin Diseases at the NIH found similar mutations on NLRP3 in patients with NOMID as well as increased expression of IL-1β mRNA in their leukocytes.[9]

## From Genetics to Therapy

In 2003, Hawkins and colleagues published the first case report of using the IL-1 receptor antagonist anakinra in patients with MWS.[21] Results were dramatic; symptoms resolved within hours of the first injection, and levels of serum amyloid A, production of which is stimulated by IL-1, normalized within 3 days.[21] Initial reports of profound activity of IL-1 inhibition in NOMID from a study starting at NIH[22] soon followed. Hoffman *et al.* demonstrated that IL-1 inhibition could prevent flares associated with cold exposure in FCAS.[23] The NIH team published their study showing that treatment with anakinra led to a rapid decrease in the signs and symptoms of NOMID, which reversed after drug withdrawal and resumed after reinitiation of therapy.[22] These studies clearly demonstrated the strong relationship between overexpression of IL-1 and symptoms in CAPS. For researchers at Regeneron, the discovery that CAPS were IL-1-driven diseases, coupled with the early success of anakinra in treating these conditions, pointed to a clear therapeutic target for rilonacept and the beginning of a collaborative development effort with colleagues in academic, government, and independent research settings.

## Rilonacept for CAPS: Initial Proof of Concept

Initial reports of anakinra in CAPS were compelling,[21–23] and a pilot study in collaboration with NIH researchers was established to test rilonacept in these conditions. This study in five patients with FCAS showed that rilonacept rapidly and effectively controlled symptoms and significantly reduced inflammatory markers.[24] A range of doses were explored and 160 mg/wk was chosen for the phase III randomized, placebo-controlled trials. However, before these studies could begin more needed to be known about the natural history of CAPS and about how to measure treatment responses.

## Monocyte/Macrophage

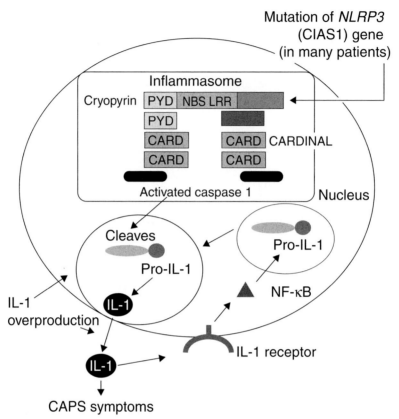

**Figure 2.** The role of the inflammasome in release of IL-1β.[17] The *NLRP3* gene encodes for cryopyrin, which forms an inflammasome with the ASC and CARDINAL adaptor proteins. This leads to activation of caspase-1 and release of IL-1β. Reproduced with permission from Hoffman, H.M. 2009. Rilonacept for the treatment of cryopyrin-associated periodic syndromes (CAPS). *Expert Opin. Biol. Ther.* **9:** 519–531. ©Informa Plc, 2009.

The most notable challenge for the clinical development of rilonacept for CAPS was the lack of a validated end point for measuring the impact of treatment. The NIH researchers had used a daily symptom diary to measure the presence and severity of symptoms in their NO-MID study, as well as the Childhood Health Assessment Questionnaire (a measure of disability) and serum levels of inflammatory markers including C-reactive protein, serum amyloid A, and erythrocyte sedimentation rate.[22]

Regeneron sought the advice of the FDA. Because of the rarity of CAPS and the lack of currently available therapies, the FDA granted rilonacept orphan drug status and priority re-view. For registration, rilonacept would need to be evaluated in a randomized, double-blind study using a validated instrument to measure symptom severity fluctuation in a manner consistent with FDA guidance on patient-reported outcomes.[25] Because of the daily fluctuations in the pattern and severity of CAPS symptoms, it was clear that a daily assessment questionnaire would be needed, similar to the one used in the NIH study,[22] but modified specifically for the less severe forms of CAPS, namely FCAS and MWS. Because of the pediatric onset of these diseases, it would also need to be suitable for both children and adults to complete. So began development of the Daily Health

Assessment Form (DHAF) in collaboration with Dr Hoffman's team at the University of California San Diego.

## Measuring Disease Activity in CAPS

Two key questions needed to be answered in the development of the DHAF:

- Does the questionnaire include the clinically relevant CAPS symptoms?
- Does the measurement scale have sufficient sensitivity to capture symptom fluctuations?

Regeneron consulted with the FDA during DHAF development to ensure that the instrument would meet regulatory requirements for a valid clinical end point. A number of versions were investigated, initially assessing the presence and severity of 11 symptoms (rash, fever, chills, joint pain, muscle pain, eye redness/pain, thirst, nausea, sweating, fatigue, and headache), and using different rating scales (5-point categorical scale and linear rating scale).[26] These versions also included questions about overall disease activity (global assessment) in different ways, and included questions about response to cold exposure, limitations of daily activities, and medication use.[26]

Because it was vital that the research produced a valid instrument, help was sought from Dr. Frederick Wolfe at the National Data Bank (NDB) for Rheumatic Diseases in Wichita to ensure the careful collection and analysis of daily questionnaires. Patients with FCAS or MWS were informed by their physician (Dr. Hoffman) of the DHAF study; interested individuals then contacted the NDB, who obtained informed consent, collected demographic and medical history information and provided the DHAF questionnaires in electronic or paper form. Questionnaires were returned online or by mail to the NDB for collation and analysis. Although the first-generation DHAF proved to have high internal consistency, two major refinements were required: The linear rating

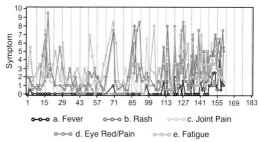

**Figure 3.** Daily DHAF symptom scores for fever, rash, joint pain, eye redness/pain and fatigue in a patient with FCAS.[26] Reproduced with permission from Hoffman H.M. *et al.* 2008. Cryopyrin-associated periodic syndromes: development of a patient-reported outcomes instrument to assess the pattern and severity of clinical disease activity. *Curr. Med. Res. Opin.* **24:** 2531–2543. ©Informa Pharmaceutical Science, 2008.

scale emerged as the preferable format for evaluating symptom severity; and fewer symptoms should be included (as suggested by the FDA).

The five symptoms which scored highest were rash, fever, joint pain, eye redness/pain and fatigue, and these showed considerable day-to-day fluctuation in severity (Fig. 3). These five symptoms were chosen for subsequent versions of the DHAF. In addition, a key symptom score (KSS) was developed as an overall assessment of symptom severity. The KSS is the sum of the scores for each symptom divided by 5 (the number of symptoms evaluated).[26] The second-generation DHAF also proved to have high internal consistency and sensitivity[26] and was included, with FDA consent, as the primary end point for the phase III trial of rilonacept in CAPS. In addition to providing a valid end point for the clinical evaluation of rilonacept, the collaborative development of the DHAF helped Regeneron to learn about the natural history of CAPS, identify potential patients for the phase III clinical trials, and perform meaningful power calculations for the placebo-controlled trials.

## Clinical Development

Two sequential placebo-controlled studies were undertaken in patients with FCAS or

MWS.[27] In the first, 47 adult patients with CAPS (44 with FCAS and 3 with MWS) were randomized to double-blind treatment with either rilonacept at a loading dose of 320 mg followed by 160 mg/wk or placebo for 6 weeks.[27] Those completing this study who were willing to participate in a further study ($n = 46$) then received 9 weeks of single-blind treatment with rilonacept 160 mg/week. After 9 weeks, patients were randomized to double-blind drug withdrawal, with 22 continuing treatment with rilonacept and 23 receiving placebo for a further nine weeks. One patient withdrew during the 9-week single-blind treatment phase.

There was a significant reduction in the KSS of patients receiving rilonacept during the initial 6-week double-blind phase, with a decrease of 84% in the rilonacept group compared with 13% in the placebo group ($P < 0.0001$) (Fig. 4A).[27] The difference between the rilonacept and placebo groups was statistically significant from the first day of treatment ($P < 0.001$).[28] Patients who switched to single-blind rilonacept at week 9 showed similar rapid improvements to those in the active treatment group in study 1 ($P < 0.0001$ vs. baseline), and symptom control was maintained in those who continued to receive rilonacept (Fig. 4A). When rilonacept was withdrawn in half the patients at week 15, symptom scores rebounded, while they were maintained at low levels in patients who continued on rilonacept (Fig. 4B). The result was a significant difference in the KSS between the rilonacept and placebo groups at the end of the evaluation period ($P = 0.0002$ at week 24).[27] This pattern of significant improvement relative to placebo was consistent across all assessment parameters: individual symptom scores ($P \leq 0.03$), physician global assessment ($P < 0.0001$), patient global assessment ($P \leq 0.003$), score for limitation of daily activities ($P \leq 0.05$), and levels of hs-CRP ($P \leq 0.0001$) and serum amyloid A ($P \leq 0.01$).[27] The number of single-symptom or multisymptom flare days was significantly reduced, from a mean of 8.6 at baseline to 0.1 single-symptom flares and from

**A**

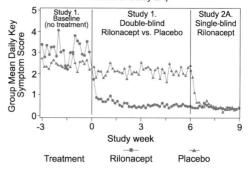

**B**

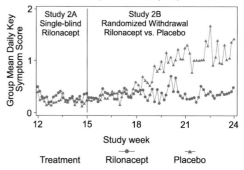

**Figure 4.** Changes in the mean key symptom score (KSS) from baseline with rilonacept and placebo during the two sequential placebo-controlled studies,[27] (**A**) at baseline, during 6 weeks of double-blind treatment with rilonacept or placebo, and during 9 weeks of single-blind rilonacept; and (**B**) during 9 weeks of single-blind rilonacept and 9 weeks of double-blind treatment with rilonacept or placebo (randomized withdrawal phase). ©*Arthritis and Rheumatism* 2008. Reproduced from Hoffman *et al.* Arthritis Rheum. 2008; 58(8): 2443–2452 with permission of Wiley-Liss, Inc. a subsidiary of John Wiley & Sons, Inc.

a mean of 13.2 to 1.1 multisymptom flares after 6 weeks of rilonacept in the initial double-blind phase.[27] Rilonacept was generally well tolerated; the most common adverse events were injection-site reactions, upper respiratory tract infections, headache, arthralgia, and diarrhea. Only one serious adverse event was reported (worsening sciatica), and this was considered unrelated to treatment.[27] Long-term open-label treatment continued, and the durability of the response was maintained for more than a year,[29,30] with no new pattern of safety or tolerability issues identified. Rilonacept was approved for use in February 2008 for adults and children aged $\geq 12$ years with FCAS or MWS.

## Beyond Orphan Indications

The rilonacept development program in CAPS revealed very useful knowledge about the drug that facilitated development in other indications. Because CAPS is a relatively homogenous disease in which genetic mutations in *NLRP3* drives IL-1β overexpression, treatment with a potent IL-1β blocker gives profound clinical responses that are detectible in small numbers of patients. Thus other diseases involving activation of the inflammasome were likely candidates for treatment with rilonacept. Moreover, we identified a highly active dosing regimen of 160 mg weekly that was effective in this overactive inflammatory disease, and would therefore likely be active in other inflammatory settings.

As part of the CAPS development program, we also devised an assay to measure the amount of IL-1β that was produced by the patients and captured by rilonacept. Although IL-1β is expressed in tissues and is cleared rapidly, it is notoriously difficult to measure in plasma. Binding to rilonacept creates an inactive complex, which accumulates because it is cleared slowly with similar pharmacokinetics to free rilonacept. Thus the level of the IL-1β:rilonacept complex at steady state reflects the relative rate of IL-1β synthesis in a given disease. We used an enzyme-linked immunosorbent assay (ELISA) incorporating a monoclonal antibody selective for human IL-1β for capture, and a biotinylated antibody against the IL-1R1 portion of rilonacept to specifically detect the complex. We found that the IL-1β:rilonacept complex levels were undetectable in normal volunteers dosed with rilonacept but were detectable in

RA patients.[31] In contrast, CAPS patients had complex levels that were on average five to sixfold higher, and in some cases more than 30-fold higher, than the mean complex level for RA patients.[31] Use of this assay therefore confirmed that IL-1β was highly overexpressed in CAPS patients, and its use could potentially reveal other disease settings where IL-1β was a driver of inflammatory disease. We thus sought to apply all of this knowledge learned from CAPS development to other larger indications.

In 2006, researchers from the University of Lausanne in Switzerland published a report demonstrating that monosodium urate crystals stimulated the inflammasome to produce IL-1β, and this mechanism contributed to the neutrophilic inflammatory response and acute symptoms of gout.[32] Furthermore, other researchers in the USA and Japan showed that the IL-1 receptor was necessary for the development of crystal-induced inflammation in gout.[33] These data pointed to a key pathogenic role for IL-1 in gout, and a potential new indication for rilonacept. This was substantiated by findings that anakinra effectively controlled gout symptoms in an open-label series of case studies in patients who had failed to respond to, or were intolerant of, standard gout therapies.[34]

In an animal model of acute gout involving administration of sodium urate crystals, administration of murine rilonacept significantly reduced inflammatory markers (levels of serum amyloid A), ankle swelling and hyperalgesia ($P < 0.05$).[35] The inflammatory response in wild-type mice after murine rilonacept administration was similar to the response in IL-1R1 knockout mice, demonstrating that IL-1 blockade was the principal mechanism for the beneficial effect. In this model, murine rilonacept was as effective in relieving ankle swelling and pain as the highest tolerated dose of colchicine, which is a less specific but established therapy for gout.[35]

With proof-of-concept established in animals, attention turned to clinical evaluation. A placebo-controlled nonrandomized monosequence pilot study was conducted in 10 patients with chronic gouty arthritis who were intolerant of, or refractory to, standard therapeutic approaches to gouty inflammation including NSAIDs, systemic or intra-articular glucocorticosteroids, or colchicine. These patients had been experiencing gout symptoms for a number of years (3–26 [mean 13 years]) and had a pain score of $\geq 3$ on a 10-point scale (mean 5.5).[36] After a 2-week, single-blind, placebo run-in phase, all patients were switched to receive subcutaneously administered single-blind rilonacept, beginning with a loading dose of 320 mg, then rilonacept 160 mg once a week for 5 weeks followed by a further 6 weeks of assessment after rilonacept withdrawal. Patients continued to take their other gout or pain medications, which could include stable doses of aspirin, nonsteroidal anti-inflammatory drugs (NSAIDs), allopurinol, probenecid, narcotics, or glucocorticoids. Rilonacept significantly reduced the median pain score ($P < 0.05$), the severity of symptoms in affected joints ($P < 0.05$) and median hs-CRP levels (59%; $P \leq 0.01$).[36] Moreover, withdrawal of rilonacept resulted in significant worsening of patients' conditions, as measured by physician and patient global assessments ($P < 0.05$), and a rebound in hs-CRP levels ($P < 0.05$).[36] We also examined the level of IL-1β:rilonacept complexes, which indicated that the chronic gouty arthritis patients expressed twice as much IL-1β compared to RA patients, but not as much as CAPS patients. These biochemical results buttressed the clinical data, and together provided justification for pursuing a phase II study in a larger population of patients with gout.

The phase II study was undertaken in patients who were initiating urate-lowering therapy for gout. Treatment initiation causes remodeling of monosodium urate crystals, which can precipitate gout flares, causing many patients to stop taking gout medication because of the early occurrence of flares.[37] A randomized, double-blind, placebo-controlled study was therefore conducted in 83 patients with gout, a serum urate level of $\geq 7.5$ mg/dL (mean 9.1 mg/dL) and $\geq 2$ symptom flares within the

preceding year. The aim of the phase II study was to determine whether rilonacept could reduce early flares associated with allopurinol therapy, which was initiated at the same time as rilonacept or placebo, and titrated from an initial dose of 300 mg to achieve a serum urate level of <6 mg/dL. Patients were not permitted to take NSAIDs or colchicine to prevent flares but could take NSAIDs or glucocorticoids for 5–10 days to treat any flares that did occur. Study treatment with allopurinol and rilonacept or placebo was continued during flares.

Compared to placebo, rilonacept significantly ($P = 0.0011$) reduced the number of gout flares patients experienced by 81%. In addition, while approximately 20% of placebo recipients experienced more than one flare, none of the rilonacept recipients experienced multiple flares. The number of days with flares and with pain scores of $\geq 5$ (on a scale of 0 to 10) were also significantly lower in the group receiving rilonacept versus placebo.[38,39] A significant difference between the treatment groups in the number of gout flares per patient was apparent from week 4 onwards ($P \leq 0.007$). These robust phase II data have led to a phase III drug development program for rilonacept in the prevention of flares induced by urate-lowering therapy as well as the treatment of acute flares.

## Conclusions

The scientific journey of rilonacept development demonstrates the interesting and unexpected paths that development of cytokine therapeutics may take. Effective collaboration between the pharmaceutical industry, academia and government agencies can produce therapeutic advances. This is particularly notable for CAPS, which are rare conditions that cause patients considerable discomfort, pain, and disability and had no effective treatment prior to the approval of rilonacept. Moreover, one can obtain detailed knowledge about the drug from its use in orphan indications, and this knowledge can be used to prioritize more common indications and streamline their development.

## Note

Some of the data from this report were presented as posters or oral presentations at a meeting entitled "Cytokine Therapies: Novel Approaches for Clinical Indications" hosted by the Food and Drug Administration and the New York Academy of Sciences in New York City on March 26–27, 2009.

## Conflicts of Interest

Drs. Stahl, Mellis, and Radin are employees of Regeneron Pharmaceuticals, Inc.

## References

1. Carpenter, L.R., G.D. Yancopoulos & N. Stahl. 1998. General mechanisms of cytokine receptor signaling. *Adv. Protein Chem.* **52:** 109–140.
2. Economides, A.N., L.R. Carpenter, J.S. Rudge, *et al.* 2003. Cytokine traps: multi-component, high-affinity blockers of cytokine action. *Nat. Med.* **9:** 47–52.
3. Dinarello, C.A. 2003. Setting the cytokine trap for autoimmunity. *Nat. Med.* **9:** 20–22.
4. Nixon, R., N. Bansback & A. Brennan. 2007. The efficacy of inhibiting tumour necrosis factor alpha and interleukin 1 in patients with rheumatoid arthritis: a meta-analysis and adjusted indirect comparisons. *Rheumatology (Oxford)* **46:** 1140–1147.
5. Garnero, P., N. Charni, N. Voorznger-Rousselot, *et al.* 2005. Effects of IL-1 Trap on biochemical markers of inflammation and joint tissue breakdown in patients with rheumatoid arthritis. *Arthritis Rheum.* **52:** S343.
6. Bingham, C.O., M. Genovese, L. Moreland, *et al.* 2004. Treatment of moderate to severe rheumatoid arthritis with IL-1 Trap [abstract]. Annual Congress of the European League Against Rheumatism. Berlin, Germany. June 9–12, 2004.
7. Bingham, C.O., M.C. Genovese, L.W. Moreland, *et al.* 2004. Results of a phase II study of IL-1 Trap in moderate to severe rheumatoid arthritis [abstract]. 68th Annual Scientific Meeting of the American College of Rheumatology. San Antonio, TX. October 16–21, 2004.
8. Guler, H.P., J.R. Caldwell, R.M. Fleischman, *et al.* 2002. Weekly treatment with IL-1 Trap is well tolerated and improved ACR criteria in patients with active rheumatoid arthritis [abstract]. Annual Congress of the European League Against Rheumatism. Stockholm, Sweden. June 12–15, 2002.
9. Aksentijevich, I., M. Nowak, M. Mallah, *et al.* 2002. De novo CIAS1 mutations, cytokine activation, and evidence for genetic heterogeneity in patients with neonatal-onset multisystem inflammatory disease (NOMID): a new member of the expanding family of pyrin-associated autoinflammatory diseases. *Arthritis Rheum.* **46:** 3340–3348.
10. Hoffman, H.M., J.L. Mueller, D.H. Broide, *et al.* 2001. Mutation of a new gene encoding a putative pyrin-like protein causes familial cold autoinflammatory syndrome and Muckle-Wells syndrome. *Nat. Genet.* **29:** 301–305.
11. Feldmann, J., A.M. Prieur, P. Quartier, *et al.* 2002. Chronic infantile neurological cutaneous and articular syndrome is caused by mutations in CIAS1, a gene highly expressed in polymorphonuclear cells and chondrocytes. *Am. J. Hum. Genet.* **71:** 198–203.
12. Hoffman, H.M. 2007. Hereditary immunologic disorders caused by pyrin and cryopyrin. *Curr. Allergy Asthma Rep.* **7:** 323–330.
13. Grateau, G. 2003. Muckle-Wells syndrome. Available at http://www.orpha.net/data/patho/GB/uk-MWS.pdf. Accessed January 12, 2009.
14. Muckle, T.J. 1979. The 'Muckle-Wells' syndrome. *Br. J. Dermatol.* **100:** 87–92.
15. Hoffman, H.M. 2003. Familial cold antoinflammatory syndrome. Available at http://www.orpha.net/data/patho/Pro/en/FamilialColdUrticaria-FRen-Pro10608.pdf. Accessed January 29, 2009.
16. Hoffman, H.M., F.A. Wright, D.H. Broide, *et al.* 2000. Identification of a locus on chromosome 1q44 for familial cold urticaria. *Am. J. Hum. Genet.* **66:** 1693–1698.
17. Hoffman, H.M. 2009. Rilonacept for the treatment of cryopyrin-associated periodic syndromes (CAPS). *Expert Opin. Biol. Ther.* **9:** 519–531.
18. Martinon, F., K. Burns & J. Tschopp. 2002. The inflammasome: a molecular platform triggering activation of inflammatory caspases and processing of proIL-beta. *Mol Cell.* **10:** 417–426.
19. Kanneganti, T.D., N. Ozoren, M. Body-Malapel, *et al.* 2006. Bacterial RNA and small antiviral compounds activate caspase-1 through cryopyrin/Nalp3. *Nature* **440:** 233–236.
20. Mariathasan, S., D.S. Weiss, K. Newton, *et al.* 2006. Cryopyrin activates the inflammasome in response to toxins and ATP. *Nature* **440:** 228–232.
21. Hawkins, P.N., H.J. Lachmann & M.F. McDermott. 2003. Interleukin-1-receptor antagonist in the

Muckle-Wells syndrome. *N. Engl. J. Med.* **348:** 2583–2584.

22. Goldbach-Mansky, R., N.J. Dailey, S.W. Canna, *et al.* 2006. Neonatal-onset multisystem inflammatory disease responsive to interleukin-1beta inhibition. *N. Engl. J. Med.* **355:** 581–592.

23. Hoffman, H.M., S. Rosengren, D.L. Boyle, *et al.* 2004. Prevention of cold-associated acute inflammation in familial cold autoinflammatory syndrome by interleukin-1 receptor antagonist. *Lancet* **364:** 1779–1785.

24. Goldbach-Mansky, R., S.D. Shroff, M. Wilson, *et al.* 2008. A pilot study to evaluate the safety and efficacy of the long-acting interleukin-1 inhibitor rilonacept (interleukin-1 Trap) in patients with familial cold autoinflammatory syndrome. *Arthritis Rheum.* **58:** 2432–2442.

25. US Department of Health and Human Services Food and Drug Administration. 2006. Guidance for Industry. Patient-reported outcome measures: use in medical product development to support labeling claims. Center for Drugs Evaluation and Research, Center for Biologics Evaluation and Research, Center for Devices and Radiological Health. Rockville, MD.

26. Hoffman, H.M., F. Wolfe, P. Belomestnov, *et al.* 2008. Cryopyrin-associated periodic syndromes: development of a patient-reported outcomes instrument to assess the pattern and severity of clinical disease activity. *Curr. Med. Res. Opin.* **24:** 2531–2543.

27. Hoffman, H.M., M.L. Throne, N.J. Amar, *et al.* 2008. Efficacy and safety of rilonacept (interleukin-1 Trap) in patients with cryopyrin-associated periodic syndromes: results from two sequential placebo-controlled studies. *Arthritis Rheum.* **58:** 2443–2452.

28. Cartwright, R., M. Throne, D. Nadler, *et al.* 2009. Rilonacept (IL-1 Trap) results in rapid and sustained reduction of clinical symptoms in patients with cryopyrin-associated periodic syndromes (CAPS) [abstract]. Cytokine Therapies: Novel Approaches to Clinical Indications. New York, NY. March 26–27, 2009.

29. Hoffman, H.M., N.J. Amar, R.C. Cartwright, *et al.* 2008. Durability of response to rilonacept (IL-1 Trap) in a phase 3 study of patients with cryopyrin-associated periodic syndromes (CAPS): Familial cold autinflammatory syndrome (FCAS) and Muckle-Wells syndrome (MWS) [abstract]. Annual Congress of the European League Against Rheumatism. Paris, France. June 11–14, 2008. *Ann. Rheum Dis.* **67**(Suppl II):104.

30. Hoffman, H.M., D. Nadler, M. Mahony, *et al.* 2009.

Durability of response to rilonacept (IL-1 Trap): Long-term follow-up in patients with cryopyrin-associated periodic syndromes (CAPS) [abstract]. Annual Congress of the European League Against Rheumatism. Copenhagen, Denmark. June 10–13, 2009. *Ann. Rheum Dis.* **68**(Suppl 3):197.

31. Torri, A., P. Belomestnov, N. Stahl, *et al.* 2009. The use of IL-1b levels as a marker for IL-1 driven diseases: comparison of IL-1b expression levels in patients with rheumatoid arthritis (RA), cryopyrin-associated periodic syndromes (CAPS) and gout [abstract]. Annual Congress of the European League Against Rheumatism. Copenhagen, Denmark. June 10–13, 2009. *Ann. Rheum Dis.* **68**(Suppl 3):189.

32. Martinon, F., V. Petrilli, A. Mayor, *et al.* 2006. Gout-associated uric acid crystals activate the NALP3 inflammasome. *Nature* **440:** 237–241.

33. Chen, C.J., Y. Shi, A. Hearn, *et al.* 2006. MyD88-dependent IL-1 receptor signaling is essential for gouty inflammation stimulated by monosodium urate crystals. *J. Clin. Invest.* **116:** 2262–2271.

34. So, A., T. De Smedt, S. Revaz, *et al.* 2007. A pilot study of IL-1 inhibition by anakinra in acute gout. *Arthritis Res. Ther.* **9:** R28.

35. Torres, R., K. Macdonald, J. Reinhardt, *et al.* 2009. Pain and inflammation in animal models of gout is reduced by m-IL-1 Trap [abstract]. Cytokine Therapies: Novel Approaches to Clinical Indications. New York, NY. March 26–27, 2009.

36. Terkeltaub, R., J. Sundy, H.R. Schumacher, *et al.* 2009. The IL-1 inhibitor rilonacept in treatment of chronic gouty arthritis: results of placebo-controlled, cross-over pilot study. *Ann. Rheum. Dis.* **68:** 1613–1617.

37. Sarawate, C.A., K.K. Brewer, W. Yang, *et al.* 2006. Gout medication treatment patterns and adherence to standards of care from a managed care perspective. *Mayo Clin. Proc.* **81:** 925–934.

38. Knapp, H.R., H.R. Schumacher, J.S. Sundy, *et al.* 2009. Rilonacept reduces the occurrence of gout flares that may be precipitated by initiation of urate-lowering therapy [abstract]. Cytokine Therapies: Novel Approaches to Clinical Indications. New York, NY. March 26–27, 2009.

39. Schumacher, H.R., J.S. Sundy, R. Terkeltaub, *et al.* 2009. Placebo-controlled study of rilonacept for prevention of gout flares during initiation of urate-lowering therapy [abstract]. Annual Congress of the European League Against Rheumatism. Copenhagen, Denmark. June 10–13, 2009. *Ann. Rheum. Dis.* **68**(Suppl 3):680.

# Actobiotics™ as a Novel Method for Cytokine Delivery

## The Interleukin-10 Case

### Lothar Steidler, Pieter Rottiers, and Bernard Coulie

*ActoGeniX NV, Zwijnaarde, Belgium*

Interleukin-10 (IL-10) is central in immune downregulation, but so far its use in inflammatory diseases remains cumbersome. For treatment of inflammatory bowel disease, adequate amounts of IL-10 must reach the intestinal lining. Systemic injection of a pharmacologically active doses of recombinant human (rh) IL-10 results in very low mucosal levels of protein and severe toxicity and side effects. In animal models, topical and active delivery of IL-10 by ingestion of recombinant *Lactococcus lactis* (*L. lactis*) was shown to be a valuable alternative. Starting thereof we have developed a novel pharmaceutical platform. Our expertise and TopAct™ (topical and active) delivery technology allows use of recombinant *L. lactis* – ActoBiotics™ – in clinical practice. Here we discuss the development of recombinant *L. lactis* for intestinal delivery of rhIL-10 in humans.

*Key words:* interleukin -10; *Lactococcus*; cytokine; therapy

## Introduction

We have pioneered genetic modification (GM) of the dairy bacterium *Lactococcus lactis* for secretion of bioactive, regulatory proteins such as cytokines and trefoil factors.[1–4] In this we saw the concept of a novel class of therapeutics: genetically modified *L. lactis* (ActoBiotics™) for topical and active (TopAct™) delivery of biologicals to the mucosa of the nose, intestine, mouth, and vagina.

Interleukin-10 (IL-10) has a clear and unique central role in establishment and maintenance of immune tolerance and in limiting the magnitude of an immune response. Inflammatory bowel disease (IBD) is characterized by lifelong chronic, relapsing inflammation within the gastrointestinal tract, driven by lack of mucosal immune tolerance towards the normal luminal microbiota.[5] IL-10 inherently holds the promise of an elegantly suited tool for treating IBD. Unfortunately, its clinical application has been encumbered by seemingly irreconcilable properties, such as very limited GI tissue penetration, an extremely low half-life, and bone marrow toxicity after systemic administration. In several mouse models of colitis (chronic dextran sulphate sodium [DSS], IL-10 deficiency, trinitrobenzene sulfonate [TNBS]), oral treatment with *L. lactis* that produce the anti-inflammatory IL-10 (*LL*IL-10) resulted in a marked reduction of the pathology, both in prophylactic as well as in a therapeutic settings.[6,7] We are developing *LL*IL-10 for the treatment of both Crohn's disease (CD) as well as ulcerative colitis (UC).

To preclude dissemination of genetically modified *L. lactis*, medical use requires an in-built environmental containment strategy.[8] Thymidylate synthase (ThyA) is an essential enzyme in DNA metabolism, providing thymine and thymidine. Genetic exchange of the chromosomal *thy*A gene for the human IL-10 gene provides *LL*IL-10 that critically depends on the addition of thymidine or thymine. Both are absent in the environment and are only present at limiting concentrations *in vivo*.

Address for correspondence: Lothar Steidler, ActoGeniX N.V., Technologiepark 4, Zwijnaarde, Belgium, 9052. Voice: +32 902610600; fax: +32 902610619. lothar.steidler@actogenix.com

Cytokine Therapies: Ann. N.Y. Acad. Sci. 1182: 135–145 (2009).
doi: 10.1111/j.1749-6632.2009.05067.x © 2009 New York Academy of Sciences.

In a Phase 1 safety and tolerability study, environmentally contained *LL*IL-10 was administered to severe Crohn's disease patients.[9] The use of *LL*IL-10 in these patients was shown to be safe, well tolerated and environmental containment was confirmed. Crohn's disease activity as well as C-reactive protein levels was clearly reduced in >50% of patients. By use of further improved *LL*IL-10, we are currently conducting a Phase 2 clinical trial in ulcerative colitis patients.

## Interleukin-10

IL-10 was initially characterized as cytokine synthesis inhibitory factor (CSIF)[10] from concanavaline A-stimulated type 2 helper T (Th2) cells. IL-10 inhibits the production of cytokines such as interleukin-2 (IL-2), tumor necrosis factor (TNF)-$\alpha$, interferon (IFN)-$\gamma$, and granulocyte macrophage colony stimulating factor (GMCSF) by Th1 cells—but not Th2 cell—in response to antigens presented by antigen-presenting cells (APC). Therefore, it has been proposed that IL-10 plays a key role in suppressing the host immune system. As elevated levels of cytokines are associated with immune diseases, IL-10 has been at the forefront of research in novel therapeutics for a multiple of immune diseases.[11]

IL-10 initiates a wide variety of activities when it binds to its cellular receptor complex. The mechanism by which IL-10 inhibits cytokine production was initially thought to be a blocking action on the antigen-presentation capacity of macrophages and dendritic cells (DC).[12-16] Continuing investigations revealed that IL-10 plays an important part not only in blocking cytokine production but also in modulating chemokine secretion and receptor expression and in the expression of various co-stimulatory molecules, including CD80, CD86, and major histocompatibility complex (MHC) Class II.[13,17-22] IL-10 also decreases beta2-integrin ligand expression (e.g., intercellular adhesion molecule 1) and decreases the generation of free radicals.[18,23] Taken together, these effects limit the activation and concurrent cytokine release by T cells during a specific immune response. IL-10 partially inhibits the induction of activities initiated by other cytokines, notably IFN-$\gamma$, IL-2, TNF-$\alpha$, and interleukin-4 (IL-4).[13]

The principle function of IL-10 therefore is to limit the magnitude of an immune response, as mice lacking IL-10 develop spontaneous enterocolitis and other symptoms akin to inflammatory bowel disease.[24] Mice lacking IL-10 following gene deletion, show increased Th1 responses and, consequentially, increased proinflammatory cytokine levels and pathologic intestinal inflammation.[25-29] These mice also show exaggerated asthmatic and allergic responses.[30-32] A recent study showed that IL-10 promotes the differentiation of CD11c$^{low}$CD45RB$^{high}$ DC that lead to the differentiation of Type 1 regulatory T cells both *in vitro* and *in vivo*, which may also explain the immunosuppressive effect of IL-10.[33] IL-10 is deep embedded in an intricately regulated mesh of counter-balancing immune messengers, which makes it less of a surprise that high dosing rather provokes undesired proinflammatory cytokine induction.[34]

## IL-10 in Inflammatory Bowel Disease

IBD comprises those conditions characterized by a tendency for chronic or relapsing immune activation and inflammation within the gastrointestinal tract. CD and UC are the two major forms of idiopathic IBD.

CD is a condition of chronic transmural granulomatous inflammation potentially involving any location of the alimentary tract from mouth to anus but with a propensity for the distal small bowel, the proximal large bowel, and the perianal region. Inflammation in CD is often discontinuous along the longitudinal axis of the gut. Affected persons usually experience the cardinal symptoms of diarrhea, abdominal pain, and, often, weight loss. Frequent complications include stricture and fistula, which often necessitate surgery.

Numerous extraintestinal manifestations may also be present. The etiology of CD is not completely understood, and therapy, although generally effective in alleviating the symptoms, is not curative. UC is a chronic, relapsing inflammatory disorder that affects the mucosal lining of the rectum and extends proximally to affect a variable extent of the colon. Clinical characteristics include rectal bleeding, diarrhea, and abdominal pain. The cause of the disease and the factors determining its chronic course are unknown.

Although the exact etiology of these diseases is not known, IBD likely results from the inappropriate and ongoing activation of the mucosal immune system driven by the presence of normal luminal flora. This aberrant response is most likely facilitated by defects in both the barrier function of the intestinal epithelium and the mucosal immune system.[5] In healthy individuals, tight regulation of the mucosal immune system prevents excessive inflammatory responses toward normal intestinal bacteria. The crucial role of IL-10 in this process is demonstrated by the observation that IL-10-deficient mice exhibit spontaneous enterocolitis.[24] Furthermore, IL-10-deficient mice kept under germfree conditions do not develop enterocolitis.[35] The latter suggests that, in the absence of immunomodulatory effects of IL-10, an unrestricted intestinal inflammatory response develops towards normal enteric antigens.

Based on the encouraging preclinical data, systemic administration of rhIL-10 has been evaluated in both CD as well as UC patients. Two multinational, multicenter studies have reported on the efficacy and safety of daily subcutaneous injections of rhIL-10 in patients with moderately active or steroid-refractory CD respectively.[36,37] Fedorak et al. reported a 24-week double-blind, placebo-controlled study in 95 patients with moderately active CD, ran-

domized to receive either daily subcutaneous administration of rhIL-10 at 1 of 4 doses (1, 5, 10, or 20 μg/kg) or placebo.[36] Following a 28-day treatment period, a modest response was seen in patients receiving 5 μg/kg rhIL-10, of whom 23.5% showed improvement compared to 0% of placebo-treated patients.

In the study reported by Schreiber *et al.*, involving 329 therapy-refractory patients with active CD, no significant differences were seen in the induction of clinical remission following administration of either rhIL-10 at 1 of 4 doses (1, 4, 8, and 20 μg/kg) or placebo. They did notice clinical improvement in 46% of 8 μg/kg rhIL-10-treated patients, compared to 27% of patients receiving placebo.[37] Schreiber *et al.* also reported on a multicenter, double-blind, placebo-controlled study of 94 patients with mild to moderately active UC, randomized to receive daily administration of either subcutaneous rhIL-10 at 1 of 4 doses (1, 5, 10, or 20 μg/kg) or placebo.[38] After a 28-day treatment period, the mean percent decrease in total ulcerative colitis scoring system (UCSS) was higher for the 5, 10, and 20 μg/kg dose levels versus placebo. However, the percent of patients achieving clinical remission was not significantly different between the active dose groups and the placebo group. Based on these data, it appears that systemic IL-10 treatment of IBD patients is only modestly effective in inducing a clinical response. Recent studies have addressed this issue and concluded that the lack of efficacy of systemic rhIL-10 is linked to:

- dose-limiting toxicity at higher systemic doses of rhIL-10,[36]
- immunostimulatory effects (induction of IFN-γ) triggered by systemic administration of rhIL-10 at higher doses,[34]
- limited tissue distribution of systemic rhIL-10 to the colonic site of inflammation,[39] and
- unfavorable pharmacokinetics of daily systemic administration of rhIL-10

(short serum half-life; between 1.1 and 2.6 hours).[40]

Nevertheless, the concept of rebalancing the intestinal immunological homeostasis with IL-10 remains very compelling and alternative methods of IL-10 delivery have been investigated. Directed local delivery of IL-10 to the mucosal site of inflammation may be more efficacious by increasing tissue bio-availability and by avoiding systemic side effects, including the proinflammatory predisposition of high doses of systemic rhIL-10.[41,42] Local therapy with rhIL-10 in an enema administration, given to three steroid-refractory UC patients for 10 days resulted in a consistent reduction in endoscopic disease score, a reduction in stool frequency and decreased secretion of proinflammatory cytokines in cultured lamina propria mononuclear cells obtained from these patients.[43] Our research showed that intragastric administration of recombinant *L. lactis* strains, engineered to secrete either murine (m)IL-10 or hIL-10, can prevent the onset of colitis in IL-10-deficient mice and significantly reduces inflammation in mouse models of colitis.[6]

## L. lactis MG1363, Source Organism for ActoBiotics

*L. lactis* is a nonpathogenic, noninvasive, noncolonizing homofermentative lactic acid bacterium and is primarily used in dairy industry. Growth can also be sustained in a select number of other nutritionally favorable areas such as specifically prepared meats and vegetable fermentations and obviously in precise laboratory culture broths. *L. lactis* cannot successfully propagate outside these very specific ecological niches. *L. lactis* MG1363 can no longer grow in milk, the primary ecological niche of its ancestor. The reason for this is twofold. In milk, the carbon source is lactose. For the fermentation/break down of lactose any bacterium will require lactose metabolizing genes, as present in the *lac* operon.

Caseins represent the amino acid source in milk. These proteins are broken down by the *PrtP* protease to provide short oligopeptides that are subsequently transported to the inside of the bacterium. *L. lactis* MG1363 is a derivative of the natural isolate, *L. lactis* NCDO 712. Gasson[44] described the removal of all of the five different plasmids that were present in NCDO 712 by protoplast-induced curing. One of those plasmids, a 33-megadalton plasmid, pLP712, was found to encode genes for lactose and casein degradation. This plasmid was required for normal growth and acid production in milk; the remaining four plasmids appeared to be cryptic. The removal of this plasmid therefore makes it impossible for MG1363 to access either one of the nutrients present in milk that are essential for its growth: lactose, providing sugars for glycolysis and caseins, providing amino acids. *L. lactis* MG1363 is therefore even more confined to specific ecological niches than its wild-type ancestor.

## ActoBiotics™, GM L. lactis for Delivery of IL-10

We have constructed *L. lactis* strains that secrete mouse (m)IL-10 (Ref. 6) and human (h)IL-10 (Ref. 8). Both proteins are expressed as a fusion with the Usp45 secretion leader.[45] During secretion, this bacterial signal sequence is cleaved off and mature IL-10 can be recovered from the growth medium. N-terminal sequencing showed correct cleavage, immediately following the bacterial secretion leader. Both proteins showed IL-10 biological activity in *in vitro* bioassays.

Whereas mIL-10 is only effective on murine cells, a limited number of other species (mouse, primates) will respond to hIL-10.[46-48] Both mIL-10- and hIL-10-secreting *L. lactis* strains can therefore be studied in murine models of colitis. In such models, intragastric administration of *LL*IL-10 successfully improved or prevented chronic colitis. Therapeutic efficacy of *LL*mIL-10 was assessed in chronic

DSS-induced colitis and in colitis found in IL-10-deficient mice and IL-10$^{-/-}$ colitis. In mice, oral administration of DSS to the drinking water induces features similar to the symptomatic and histological findings in human chronic colitis: signs of erosion, prominent regenerations of the colonic mucosa (including dysplasia), shortening of the large intestine, and frequent formation of lymphoid follicles. IL-10-deficient mice, generated by gene targeting, show normal lymphocyte development and antibody responses, but suffer from chronic enterocolitis.[24] Intragastric administration of $2 \times 10^7$ CFU *LL*mIL-10 daily for 2 (DSS mice) or 4 weeks (IL10-deficient mice) respectively, efficiently cured colitis in the chronic DSS model and prevented the onset of colitis in IL-10-deficient mice.[6]

In DSS-induced colitic mice, administration of *LL*mIL-10 was shown to be as effective as positive controls, that is, treatment with dexamethasone, anti-IL-12 antibodies or systemically administered mIL-10: all therapies decreased inflammation in DSS-induced colitis significantly, but administration of *LL*mIL-10 induced this therapeutic effect at a dose that was approximately 10,000-fold lower than the dose for systemically administered mIL-10 (Fig. 1). Ultraviolet light (UV) killing of the *LL*mIL-10 prior to inoculation abrogated the therapeutic effect, indicating that the therapeutic effects required physiologically active *LL*mIL-10 (Fig. 1). Moreover, mouse adoptive transfer experiments have shown that *L. lactis*-mediated IL-10 delivery leads to the induction of a CD4$^+$ T-cell compartment which suppresses the activity of colitic T cells, contributing to the therapeutic effect of the engineered *L. lactis* strains (our unpublished data)

To provide proof of concept that indeed also *LL*hIL-10 can be used to treat intestinal immune pathology, we used anti-CD40-induced colitis.[49] Intragastric administration of *LL*hIL-10 at different doses, revealed that *LL*hIL-10 at $10^{10}$ CFU, $10^9$ CFU, and $10^7$ CFU daily resulted in profound improvement in colitis, reduction of serum IFN-$\gamma$ and IL-6, mu-cosal MCP-1, IFN-$\gamma$ and IL-6, and a reduction in the number and activation of DC in the mesenteric lymph nodes (our unpublished data).

## Pharmacokinetic Profile of ActoBiotics™

*L. lactis* does not integrate within the microbiota that inhabits the mammalian intestine. Upon ingestion it passes through at the rate of an inert marker and disappears from the human GI tract.[50] In rats, *L. lactis* cells are quite resistant to stomach passage but only 10 to 30% will survive passage through the duodenum.[51] Despite their inability to colonize, live *L. lactis* remain metabolically active inside the intestine.[51,52]

With respect to *LL*IL-10 we have confirmed these findings, both in healthy as well as colitic mice. Up to 4 hours after gavage, *LL*IL-10 can be found back in all segments of the intestine. The numbers of CFU recovered add up to approximately 10%. Peak recovery in the feces lies around 6–8 hours and diminishes exponentially thereafter. The presence of *LL*hIL-10 coincides with the presence of hIL-10, the latter being restricted to the intestinal wall. No hIL-10 could be detected in the mesenteric lymph nodes or in systemic circulation. In inflamed intestine *L. lactis* cells were found in the submucosa but deeper penetration was never observed by ourselves as well as by others.[53] Also here we could not measure systemic availability of hIL-10. Most interestingly however, systemic immune effects have been observed following *L. lactis* mediated delivery of mIL-10[54] or other cargo proteins[55] to the healthy intestine. The mechanism for this involves the local induction of regulatory T cells.[55] It may also involve sampling by dendritic cells through transepithelial dendrites. These were occasionally observed close to luminal *Lactococci* and *L. lactis* have been observed inside phagocytosing cells.[53]

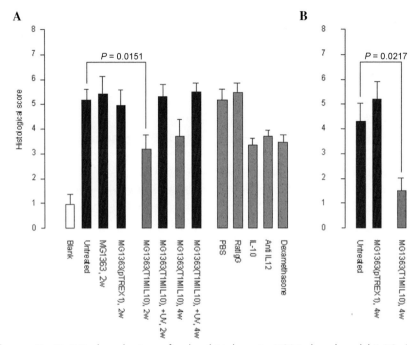

**Figure 1.** Statistical evaluation of colon histology in DSS-induced and hIL-10$^{-/-}$ colitic mice. Colon sections were randomly numbered and interpreted in a blinded manner. Scores from individual mice were subsequently decoded, and regrouped numbers were analyzed statistically. Bars represent the mean ± SEM. *$P < 0.025$; **$P = 0.0151$. (**A**) Histological scores (sum of epithelial damage and lymphoid infiltrate, both ranging from 0 to 4) for the distal colon of groups ($n = 10$) of control female Balb/c mice (white bar) and of female Balb/c mice with DSS-induced colitis that were untreated, treated with the indicated *L. lactis* cultures (black and red bar), or treated with five daily intraperitoneal injections of the compounds indicated (blue bars) (mIL-10: 5 μg per mouse per day; anti-IL-12 monoclonal antibodies: 1 mg per mouse per day; dexamethasone: 5 μg per mouse per day; rat IgG: 5 μg per mouse per day). Mice treated daily for 2 or 4 weeks with $2 \times 10^7$ CFU mIL-10-producing MG1363[T1mIL10] showed significantly reduced inflammation when compared with untreated or control-treated (unmodified MG1363 or empty vector MG1363[pTREX1]) mice. This effect was not observed when MG1363[T1mIL10] cultures were UV-killed (+UV). (**B**) Histological scores (sum of the degrees of inflammation in the proximal, middle, and distal colon, all ranging between 0 and 4) obtained after blinded interpretation of groups ($n = 5$) of 7-week-old untreated, MG1363[pTREX1]-treated (black bar), or MG1363[T1mIL10]-treated (red bar) female IL-10-deficient mice. MG1363[T1mIL10]-treated mice showed significantly less inflammation than untreated mice.

## Environmental Containment and Genetic Stability

GM *L. lactis*, engineered to secrete hIL-10, might provide a new therapeutic approach for IBD. However, the release of GMOs through clinical use raises legitimate concerns on the deliberate release of genetically modified organisms. These concerns relate to the dissemination of antibiotic selection markers that are present on expression plasmids, the survival and propagation of the GMO in the environment and the dissemination of the genetic modification to other micro-organisms. We were the first to describe environmental containment by replacing *thy*A by a "cargo" gene.[8] In this approach, the chromosomally located *thy*A gene is replaced by the hIL-10 (cargo) gene. Propagation of the hIL-10 gene now occurs along with the bacterial chromosome and does not require a plasmid. In order to control dissemination of therapeutic GM microflora, environmental containment through the use of *thy*A-deficient *L. lactis* is gradually becoming the accepted standard resulting in approvals by the Belgian Biosafety Advisory Council, the Swedish Medical Products Agency, Health Canada, and the Canadian Environmental Protection Agency. Not only are *thy*A-deficient strains critically dependent on thymidine supplementation for growth and survival, chromosomal location of the GM trait also provides superior stability throughout generations. We tested siblings of ActoBiotics™ separated over 100 generations and could not observe any difference in their structure. Chromosomal localization of the transgenic trait further obviates concerns on horizontal gene transfer.

## Safety, a Key Hallmark of ActoBiotics™

For any pharmaceutical preparation, safety is of paramount importance. To a large extent we chose to develop the food grade bacterium *L. lactis* because of its intrinsic harmless nature. Initially, we relied on state-of-the-art knowledge around food products obtained through *L. lactis* fermentation. An extended history of safe use has taught that all healthy adults, elderly, and infants but also diseased and immune compromised people can consume large amounts of live *L. lactis* without experiencing any health problems. The literature contains only very scarce examples of (mostly opportunistic) infections with *L. lactis*.

Through our experimentation, we have had the opportunity to observe numerous pathological parameters in diseased settings and never have we observed worsening caused by *L. lactis*. Our pharmacodynamic studies show that, upon oral gavage, *L. lactis* cells are not retained inside animals, cannot be recovered from blood and, even in colitic animals, do not disseminate outside of the gut mucosa. We further found that *L. lactis* cause no infections, even not

in immune deficient animals. We inoculated $10^6$ *thy*A-deficient *L. lactis* in severely immune compromised, neutropenic rats (rendered neutropenic by injection with cyclophosphamide). Bacteria were found back in major peripheral tissues shortly after injection but 6 hours post dosing all *L. lactis* had disappeared from the animals. Also in these animals, *L. lactis* were never found back in the circulation, which is in good agreement with our findings that *L. lactis* cannot survive in human serum.

According to the ICH (The International Conference on Harmonisation of Technical Requirements for Registration of Pharmaceuticals for Human Use) guidelines, we conducted toxicity studies in mice and in primates. A subchronic 13-week toxicity study of *LL*IL-10 in mice, including a 6-week recovery period showed that twice daily administration was not associated with any treatment-related signs of toxicity or adverse effects. No evidence was found for anti-hIL-10 antibodies in plasma samples. Following this, a 4-week repeat-dose toxicity study including a 4-week recovery period, combining oral (gavage) and rectal administration of *LL*hIL-10, was performed in cynomolgus monkeys. This showed that twice daily combined oral and rectal treatment of monkeys with *LL*hIL-10 for 4 consecutive weeks was not associated with any treatment-related signs of toxicity or adverse effects. Both toxicology studies showed that the no observed adverse effects level (NOAEL) of *LL*hIM-10 equals the highest dose given.

## Phase I Study in Crohn's Disease Patients

Using *LL*hIL-10 (product code Thy12), a Phase 1 open label, placebo-uncontrolled safety and tolerability study was conducted in which 10 CD patients were enrolled.[56] The treatment period was seven days, and patients were assessed daily for the presence of potential adverse effects by direct questioning and assessment of disease activity. The presence and

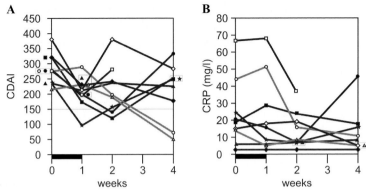

**Figure 2.** Clinical scores of patients before and after Thy12 administration. Crohn's disease activity index (CDAI; A) and C-reactive protein levels (CRP; B) in patients at weeks 0, 1, 2, and 4. The black bar in the time axis indicates treatment period of 7 days. Boxed symbols represent identical values for different patients. During treatment, CDAI decreases for all patients except two (gray lines) who however show a substantial decrease afterwards. In five patients, CDAI decreased below 150 (generally regarded as cut-off value for remission induction). CRP showed a tendency of decrease especially in patients with high CRP values.

kinetics of Thy12-release in the stool of patients was determined by conventional culturing and quantitative polymerase chain reaction (Q-PCR) of Thy12 gene sequences. CDAI score and serum levels of C-reactive protein (CRP) were measured at days −1, 8, 14, and 28. Patients had a mean decrease in CDAI of 71.7 after 1 week of treatment, the decrease in soft stool being a prominent factor in this, and showed a concomitant decrease in CRP levels (Fig. 2). Based on the pooled data from all time points, a clinical benefit was observed in 8 of 10 patients; five patients went into complete clinical remission (CDAI, <150), and three patients exhibited a clinical response (decrease in CDAI, >70). One patient did not respond, and one was excluded from the analysis because of withdrawal from the study (noncompliance with the dosing regimen and persistent vomiting).

Overall, during treatment with Thy12, no serious adverse events were reported. A decrease in disease activity was observed. Fecally-recovered Thy12 bacteria were dependent on thymidine for growth and showed IL-10 production, indicating that the GM traits were stably maintained. Additionally, the environmental containment strategy was validated *in vivo*.

## Discussion

ActoBiotics[TM] have successfully passed preclinical development stages and subsequent stringent review by the Regulatory Authorities resulting in both clinical as well as environmental approval. Functionality of *LL*IL-10 has been shown in a multitude of mouse models by ourselves (extensively described above) as well as by others.[7,54,57] The encouraging results of the Phase 1 study led to further clinical development of *LL*IL-10. Clinical application however requires much broader documentation than mere proof of concept studies in animals and technology of preparation must be reliable and robust to guarantee, in all senses, a "viable product."

In the food industry, production technology is well established for *L. lactis*. Off-the-shelf technology and know-how for pharmaceutical grade cGMP (current good manufacturing practice) production and formulation of *L. lactis* was however not available. We have set up a streamlined procedure for the manufacturing of bulk amounts of *LL*IL-10. We now have a reliable, robust, and transferable production process to allow clinical experimentation. In general, it consists of growing *L. lactis* to high density, freeze drying, and milling to obtain a dry powder. Formulations have been prepared in which the *L. lactis* powder is put in capsules with enteric coating but also enema and oral rinse dosage forms are available. We have put in place the appropriate specifications of purity, identity, dosage, and strength and have established standard operating procedures to measure them. This enables us to follow the key characteristics of the product over time, showing excellent product stability.

We have solved a great number of potential issues and are now ready to thoroughly evaluate ActoBiotics™. The availability of all aspects of a completely novel technology allows us to conduct a phase 2A randomized, double-blind, placebo controlled, multicenter dose escalation clinical trial in mild to moderate ulcerative colitis patients. During 28 days, 40 patients will now receive *LL*IL-10 preparation and 20 will receive placebo, both as a capsule as well as an enema formulation. Three dosage levels are tested and end points are safety, tolerability, pharmacodynamic effect, and efficacy of *LL*IL-10.

If anything is striking in TopAct™ technology, it is the versatility of the platform. We have established many of the essential features of this technology and found that many aspects can be easily translated to other products: *L. lactis* strains that are genetically engineered to express other therapeutic proteins. Therefore we are convinced that TopAct™ has the potential to become a broadly applicable pharmaceutical platform.

### Competing Interests

Disclosure of competing interests: LS, PR and BC are employees of AcoGeniX N.V; LS and BC are shareholders of ActoGeniX N.V.

TopAct™ and ActoBiotics™ are trademarks registered by ActoGeniX.

# References

1. Schotte, L., L. Steidler, J. Vandekerckhove, *et al.* 2000. Secretion of biologically active murine interleukin-10 by Lactococcus lactis. *Enzyme Microbial. Technol.* **27:** 761–765.

2. Steidler, L., K. Robinson, L. Chamberlain, *et al.* 1998. Mucosal delivery of murine interleukin-2 (IL-2) and IL-6 by recombinant strains of Lactococcus lactis coexpressing antigen and cytokine. *Infect. Immun.* **66:** 3183–3189.

3. Steidler, L., J.M. Wells, A. Raeymaekers, *et al.* 1995. Secretion of biologically active murine interleukin-2 by Lactococcus lactis subsp. lactis. *Appl. Environ. Microbiol.* **61:** 1627–1629.

4. Vandenbroucke, K., W. Hans, J. Van Huysse, *et al.* 2004. Active delivery of trefoil factors by genetically modified Lactococcus lactis prevents and heals acute colitis in mice. *Gastroenterology* **127:** 502–513.

5. Podolsky, D.K. 2002. Inflammatory bowel disease. *N. Engl. J. Med.* **347:** 417–429.

6. Steidler, L., W. Hans, L. Schotte, *et al.* 2000. Treatment of murine colitis by Lactococcus lactis secreting interleukin-10. *Science* **289:** 1352–1355.

7. Foligne, B., S. Nutten, L. Steidler, *et al.* 2006. Recommendations for improved use of the murine TNBS-induced colitis model in evaluating anti-inflammatory properties of lactic acid bacteria: Technical and microbiological aspects. *Dig. Dis. Sci.* **51:** 390–400.

8. Steidler, L., S. Neirynck, N. Huyghebaert, *et al.* 2003. Biological containment of genetically modified Lactococcus lactis for intestinal delivery of human interleukin 10. *Nat. Biotechnol.* **21:** 785–789.

9. Braat, H., P. Rottiers, N. Huyghebaert, *et al.* 2006. IL-10 producing Lactococcus lactis for the treatment of Crohn's disease. *Inflamm. Bowel Dis.* **12:** S26–S27.

10. Fiorentino, D.F., M.W. Bond & T.R. Mosmann. 1989. Two types of mouse T helper cell. IV. Th2 clones secrete a factor that inhibits cytokine production by Th1 clones. *J. Exp. Med.* **170:** 2081–2095.

11. Pestka, S., C.D. Krause, D. Sarkar, *et al.* 2004. Interleukin-10 and related cytokines and receptors. *Annu. Rev. Immunol.* **22:** 929–979.

12. Ding, L. & E.M. Shevach. 1992. IL-10 inhibits mitogen-induced T cell proliferation by selectively inhibiting macrophage costimulatory function. *J. Immunol.* **148:** 3133–3139.

13. Fiorentino, D.F., A. Zlotnik, P. Vieira, *et al.* 1991. IL-10 acts on the antigen-presenting cell to inhibit cytokine production by Th1 cells. *J. Immunol.* **146:** 3444–3451.

14. Macatonia, S.E., T.M. Doherty, S.C. Knight, *et al.* 1993. Differential effect of IL-10 on dendritic cell-induced T cell proliferation and IFN-gamma production. *J. Immunol.* **150:** 3755–3765.

15. Enk, A.H., V.L. Angeloni, M.C. Udey, *et al.* 1993. Inhibition of Langerhans cell antigen-presenting

function by IL-10. A role for IL-10 in induction of tolerance. *J. Immunol.* **151:** 2390–2398.

16. Peguet-Navarro, J., C. Moulon, C. Caux, *et al.* 1994. Interleukin-10 inhibits the primary allogeneic T cell response to human epidermal Langerhans cells. *Eur. J. Immunol.* **24:** 884–891.

17. Ding, L., P.S. Linsley, L.Y. Huang, *et al.* 1993. IL-10 inhibits macrophage costimulatory activity by selectively inhibiting the up-regulation of B7 expression. *J. Immunol.* **151:** 1224–1234.

18. Willems, F., A. Marchant, J.P. Delville, *et al.* 1994. Interleukin-10 inhibits B7 and intercellular adhesion molecule-1 expression on human monocytes. *Eur. J. Immunol.* **24:** 1007–1009.

19. Jinquan, T., C.G. Larsen, B. Gesser, *et al.* 1993. Human IL-10 is a chemoattractant for CD8+ T lymphocytes and an inhibitor of IL-8-induced CD4+ T lymphocyte migration. *J. Immunol.* **151:** 4545–4551.

20. Kasama, T., R.M. Strieter, N.W. Lukacs, *et al.* 1994. Regulation of neutrophil-derived chemokine expression by IL-10. *J. Immunol.* **152:** 3559–3569.

21. Sozzani, S., S. Ghezzi, G. Iannolo, *et al.* 1998. Interleukin 10 increases CCR5 expression and HIV infection in human monocytes. *J. Exp. Med.* **187:** 439–444.

22. Takayama, T., A.E. Morelli, N. Onai, *et al.* 2001. Mammalian and viral IL-10 enhance C-C chemokine receptor 5 but down-regulate C-C chemokine receptor 7 expression by myeloid dendritic cells: impact on chemotactic responses and in vivo homing ability. *J. Immunol.* **166:** 7136–7143.

23. Dokka, S., X. Shi, S. Leonard, *et al.* 2001. Interleukin-10-mediated inhibition of free radical generation in macrophages. *Am. J. Physiol. Lung. Cell. Mol. Physiol.* **280:** L1196–L1202.

24. Kuhn, R., J. Lohler, D. Rennick, *et al.* 1993. Interleukin-10-deficient mice develop chronic enterocolitis. *Cell* **75:** 263–274.

25. Yang, X., J. Gartner, L. Zhu, *et al.* 1999. IL-10 gene knockout mice show enhanced Th1-like protective immunity and absent granuloma formation following Chlamydia trachomatis lung infection. *J. Immunol.* **162:** 1010–1017.

26. Murray, P.J. & R.A. Young. 1999. Increased antimycobacterial immunity in interleukin-10-deficient mice. *Infect. Immun.* **67:** 3087–3095.

27. Sewnath, M.E., D.P. Olszyna, R. Birjmohun, *et al.* 2001. IL-10-deficient mice demonstrate multiple organ failure and increased mortality during Escherichia coli peritonitis despite an accelerated bacterial clearance. *J. Immunol.* **166:** 6323–6331.

28. Vazquez-Torres, A., J. Jones-Carson, R.D. Wagner, *et al.* 1999. Early resistance of interleukin-10 knockout mice to acute systemic candidiasis. *Infect. Immun.* **67:** 670–674.

29. Gazzinelli, R.T., M. Wysocka, S. Hieny, *et al.* 1996. In the absence of endogenous IL-10, mice acutely infected with Toxoplasma gondii succumb to a lethal immune response dependent on CD4+ T cells and accompanied by overproduction of IL-12, IFN-gamma and TNF-alpha. *J. Immunol.* **157:** 798–805.

30. Tournoy, K.G., J.C. Kips & R.A. Pauwels. 2000. Endogenous interleukin-10 suppresses allergen-induced airway inflammation and nonspecific airway responsiveness. *Clin. Exp. Allergy.* **30:** 775–783.

31. Berg, D.J., M.W. Leach, R. Kuhn, *et al.* 1995. Interleukin 10 but not interleukin 4 is a natural suppressant of cutaneous inflammatory responses. *J. Exp. Med.* **182:** 99–108.

32. Grunig, G., D.B. Corry, M.W. Leach, *et al.* 1997. Interleukin-10 is a natural suppressor of cytokine production and inflammation in a murine model of allergic bronchopulmonary aspergillosis. *J. Exp. Med.* **185:** 1089–1099.

33. Wakkach, A., N. Fournier, V. Brun, *et al.* 2003. Characterization of dendritic cells that induce tolerance and T regulatory 1 cell differentiation in vivo. *Immunity* **18:** 605–617.

34. Tilg, H., C. van Montfrans, A. van den Ende, *et al.* 2002. Treatment of Crohn's disease with recombinant human interleukin 10 induces the proinflammatory cytokine interferon gamma. *Gut.* **50:** 191–195.

35. Sellon, R.K., S. Tonkonogy, M. Schultz, *et al.* 1998. Resident enteric bacteria are necessary for development of spontaneous colitis and immune system activation in interleukin-10-deficient mice. *Infect Immun.* **66:** 5224–5231.

36. Fedorak, R.N., A. Gangl, C.O. Elson, *et al.* 2000. Recombinant human interleukin 10 in the treatment of patients with mild to moderately active Crohn's disease. The Interleukin 10 Inflammatory Bowel Disease Cooperative Study Group. *Gastroenterology* **119:** 1473–1482.

37. Schreiber, S., R.N. Fedorak, O.H. Nielsen, *et al.* 2000. Safety and efficacy of recombinant human interleukin 10 in chronic active Crohn's disease. Crohn's Disease IL-10 Cooperative Study Group. *Gastroenterology* **119:** 1461–1472.

38. Schreiber, S., R.N. Fedorak, G. Wild, *et al.* 1998. Safety and tolerance of rhIL-10 treatment in patients with mild/moderate active ulcerative colitis. *Gastroenterology* **114:** 1080.

39. Rachmawati, H., L. Beljaars, C. Reker-Smit, *et al.* 2004. Pharmacokinetic and biodistribution profile of recombinant human interleukin-10 following intravenous administration in rats with extensive liver fibrosis. *Pharm. Res.* **21:** 2072–2078.

40. van Deventer, S.J., C.O. Elson & R.N. Fedorak. 1997. Multiple doses of intravenous interleukin 10 in steroid-refractory Crohn's disease. Crohn's Disease Study Group. *Gastroenterology* **113:** 383–389.

41. Herfarth, H. & J. Scholmerich. 2002. IL-10 therapy in Crohn's disease: at the crossroads. Treatment of Crohn's disease with the anti-inflammatory cytokine interleukin 10. *Gut.* **50:** 146–147.

42. Madsen, K. 2002. Combining T cells and IL-10: a new therapy for Crohn's disease? *Gastroenterology* **123:** 2140–2144.

43. Schreiber, S., T. Heinig, H.G. Thiele, *et al.* 1995. Immunoregulatory role of interleukin 10 in patients with inflammatory bowel disease. *Gastroenterology* **108:** 1434–1444.

44. Gasson, M.J. 1983. Plasmid complements of Streptococcus lactis NCDO 712 and other lactic streptococci after protoplast-induced curing. *J. Bacteriol.* **154:** 1–9.

45. van Asseldonk, M., G. Rutten, M. Oteman, *et al.* 1990. Cloning of usp45, a gene encoding a secreted protein from Lactococcus lactis subsp. lactis MG1363. *Gene.* **95:** 155–160.

46. Moore, K.W., R. de Waal Malefyt, R.L. Coffman, *et al.* 2001. Interleukin-10 and the interleukin-10 receptor. *Annu. Rev. Immunol.* **19:** 683–765.

47. Rosenblum, I.Y., R.C. Johnson & T.J. Schmahai. 2002. Preclinical safety evaluation of recombinant human interleukin-10. *Regul. Toxicol. Pharmacol.* **35:** 56–71.

48. Termont, S., K. Vandenbroucke, D. Iserentant, *et al.* 2006. Intracellular accumulation of trehalose protects Lactococcus lactis from freeze-drying damage and bile toxicity and increases gastric acid resistance. *Appl. Environ. Microbiol.* **72:** 7694–7700.

49. Uhlig, H.H., B.S. McKenzie, S. Hue, *et al.* 2006. Differential activity of IL-12 and IL-23 in mucosal and systemic innate immune pathology. *Immunity* **25:** 309–318.

50. Klijn, N., A.H. Weerkamp & W.M. de Vos. 1995. Genetic marking of Lactococcus lactis shows its survival in the human gastrointestinal tract. *Appl. Environ. Microbiol.* **61:** 2771–2774.

51. Drouault, S., G. Corthier, S.D. Ehrlich, *et al.* 1999. Survival, physiology, and lysis of Lactococcus lactis in the digestive tract. *Appl. Environ. Microbiol.* **65:** 4881–4886.

52. Roy, K., J. Anba, G. Corthier, *et al.* 2008. Metabolic adaptation of Lactococcus lactis in the digestive tract: the example of response to lactose. *J. Mol. Microbiol. Biotechnol.* **14:** 137–144.

53. Waeytens A, L. Ferdinande, S. Neirynck, *et al.* 2008. Paracellular entry of interleukin-10 producing Lactococcus lactis in inflamed intestinal mucosa in mice. *Inflamm. Bowel Dis.* **14:** 471–479.

54. Frossard, C.P., L. Steidler & P.A. Eigenmann. 2007. Oral administration of an IL-10-secreting Lactococcus lactis strain prevents food-induced IgE sensitization. *J. Allergy Clin. Immunol.* **119:** 952–959.

55. Huibregtse, I.L., V. Snoeck, A. de Creus, *et al.* 2007. Induction of ovalbumin-specific tolerance by oral administration of Lactococcus lactis secreting ovalbumin. *Gastroenterology* **133:** 517–528.

56. Braat, H., P. Rottiers, D.W. Hommes, *et al.* 2006. A phase I trial with transgenic bacteria expressing Interleukin-10 in Crohn's disease. *Clin. Gastroenterol. Hepatol.* **4:** 754–759.

57. Foligne, B., R. Dessein, M. Marceau, *et al.* 2007. Prevention and treatment of colitis with Lactococcus lactis secreting the immunomodulatory Yersinia LcrV protein. *Gastroenterology* **133:** 862–874.

# Hurdles and Leaps for Protein Therapeutics

## Cytokines and Inflammation

**Steven Kozlowski,[a] Barry Cherney,[b] and Raymond P. Donnelly[b]**

[a]*Office of Biotechnology Products, Office of Pharmaceutical Science, Center for Drug Evaluation and Research, U.S. FDA, Silver Spring, Maryland, USA*

[b]*Division of Therapeutic Proteins, Office of Biotechnology Products, Office of Pharmaceutical Science, Center for Drug Evaluation and Research, U.S. FDA, Bethesda, Maryland, USA*

Cytokines encompass a wide variety of proteins that can trigger many cellular activities. An important set of cytokines modulate inflammatory responses (inflammatory cytokines). These molecules have potent biological activities and have been a major focus for protein drug development. There have been both successes and failures in this area. Initial hurdles, such as limited manufacturing capacity, have now been largely overcome. However clinical development remains a challenge. On the basis of the history of cytokine therapeutics, a number of strategies for future drug development are considered.

*Key words:* cytokines; biotechnology; inflammation; pharmaceutical development; manufacturing

## Introduction

Cytokines are signaling molecules that are used extensively in cellular communication. Cytokines that impact the activities of the immune system have been studied for more than 50 years. Interferons (IFN) were discovered in 1957 and have been of great interest to the scientific community ever since.[1] These molecules induce a broad range of biological activities. The potential clinical benefits from these potent biological effects generated a lot of excitement. Publicity on IFN clinical studies in the late 1970s led to high expectations for cancer treatments.[1,2] Although IFN did not live up to all these expectations, this group of cytokines has a variety of U.S. FDA approved products marketed for a number of clinical indications, including chronic hepatitis C virus infection and several types of cancer.

The genes for interleukin 1 (IL-1) alpha and beta were cloned in 1984.[3] However, for decades before this, the biological activities of these gene products were studied as endogenous pyrogen and leukocytic endogenous mediator.[4] Despite the importance of these molecules in inflammatory responses and pathology, it took until 2001 for an IL-1 receptor antagonist to be approved for rheumatoid arthritis[5] and another seven years for the approval of additional orphan indications for another IL-1 antagonist.[6]

The proper matching of biological activities to disease states for these cytokines evolved over decades and involved many scientists in academia and industry. Currently, there are dozens of known cytokines and advances in genomics and receptor-ligand pairing[7] may continue to increase the numbers of known cytokines. Most of these molecules have potent biological activities leading to high expectations for clinical utility. Clinical setbacks with some of

Address for correspondence: Dr. Steven Kozlowski, Office of Biotechnology Products, OPS, CDER, Food and Drug Administration, White Oak Building 21 Room 1510, 10903 New Hampshire Ave, Silver Spring, MD 20993. Voice: 301-796-2046; fax: 301-796-9743. steven.kozlowski@fda.hhs.gov

Cytokine Therapies: Ann. N.Y. Acad. Sci. 1182: 146–160 (2009).
doi: 10.1111/j.1749-6632.2009.05158.x © 2009 New York Academy of Sciences.

these products, such as IL-12 and IL-10, have slowed development of this class of drugs.[8,9] The powerful impact of cytokines on biology suggests that an initial cytokine product failure should not necessarily trigger product abandonment. Setbacks are often a part of pharmaceutical development, and therapeutic proteins have faced many hurdles over the past decades. Innovation and perseverance have overcome these hurdles. The hurdles of therapeutic protein development and the strategies that leapt over them are models for enhancing the development of an important subset of therapeutic proteins, cytokines. Some of these historical hurdles and solutions are discussed below. Some current challenges for cytokine development discussed at the NYAS/FDA meeting on this topic will also be considered.[10]

## Product Supply

### Source Material

Proteins were initially extracted from tissues or body fluids. Interferon alphas used were generated by stimulating human leukocytes.[11] Natural source material could have safety concerns, such as transmission of Jakob-Creutzfeldt disease from pituitary-derived growth hormone.[12] There were also limits to the amount of donor tissues available. In the early years of cytokine drug development, it was often challenging to generate sufficient material for clinical trials and even more so to generate enough material for marketing. Genetic engineering provided a solution to the limitations of natural source materials. Despite the clear promise of such an approach, the use of this technology in manufacture of protein products was a gradual and deliberate process.

In the early 1950s, DNA was shown to transfer genetic information, and the double helix structure of DNA was resolved. By 1961, the genetic code was deciphered opening the way to genetic engineering and biotechnology. Recombination allowed insertion of genes encoding therapeutic proteins into host cells, such as bacteria. These advances promised tremendous benefits, but also generated a great deal of unnecessary fear. In 1976, the NIH issued a guideline on the use of recombinant DNA to deal with the potential safety risks of this new technology.[13] The following year, there were many bills before the U.S. Congress to limit the use of recombinant DNA technology. Between 1976 and 1986 this fear diminished and genetic engineering allowed for the successful production and marketing of human insulin and growth hormone which were manufactured using recombinant bacterial expression systems. Generation of sufficient amounts of IFN was also enabled through the use of a recombinant source.[14]

Although recombinant bacterial source materials have been of great value, they have limitations, such as the inability to glycosylate proteins when expressed in bacteria such as *E. coli*. The murine monoclonal antibody, Muromonab-CD3, was the first marketed product made from an immortalized mammalian cell fusion.[15] Shortly thereafter, alteplase was approved for treatment of acute myocardial infarctions. Alteplase production uses immortalized Chinese hamster ovarian cells,[16] further advancing the role of mammalian cell lines in the manufacture of biotechnology products. A granulocyte macrophage-colony stimulating factor (GM-CSF) was produced in recombinant yeast, expanding the range of host organisms for production of bioactive proteins, including cytokines. Currently, baculovirus vectors in insect cells are being used to express proteins for candidate vaccines.[17] In February 2009, an anticoagulant protein derived from the milk of transgenic goats was approved in the United States.[18] Other novel organisms, such as transgenic plants, are being developed as hosts for production of recombinant proteins.[19]

### Manufacturing Flexibility

Issues other than source materials can also limit the production of therapeutic proteins.

Protein products are complex and have many structural attributes. They may contain a variety of potential impurities. Although impacted by source material, protein structural attributes and the level of impurities can also be affected by the manufacturing process.

Since end-product testing was of very limited utility in assessing clinical performance of complex biological products, the manufacturing process was critical to assurance of consistent product quality. The strategy to deal with this process dependence was to fix the commercial manufacturing process to the process used for manufacture of clinical lots. Thus, the product is linked to the labeled clinical outcomes through a highly defined manufacturing process. The biologics mantra of "the process is the product" was coined to describe this linkage. This limited flexibility in manufacturing for a successful product. It also limited clinical studies by requiring the final commercial process be in place for the product lots used in the pivotal clinical trials. However, the need for efficient product development and the capacity to respond to variable market demands drove development of a more flexible approach to deal with manufacturing changes.

Although end-product testing still cannot fully define the structure and impurity profiles of most protein products, the characterization of proteins has greatly advanced over time. Initially the biochemical characterization of proteins was limited to tools such as gel electrophoresis and chromatographic analysis. Peptide mapping[20] and carbohydrate analysis improved characterization. The capabilities of mass spectroscopy and nuclear magnetic resonance have grown rapidly and enhanced protein analytics. The pioneers of these analytical techniques for biological macromolecules shared a Nobel Prize for chemistry.[21] Although bioassays have always been a part of protein characterization, improved bioassays with lower variances and greater high throughput potential have been developed.

On the basis of these technological advances, it was felt that the comparative characterization of pre- and postchange product could allow some manufacturing changes with limited or no clinical data. The concept that product characterization may allow for process changes was integrated into documents on comparability.[22,23] In 1997, a new approach to reporting of manufacturing changes was established through the Food and Drug Administration Modernization Act. Regulations on the reporting categories for biologics manufacturing changes were fleshed out in guidance.[24] Other guidance by the U.S. FDA[25] and the International Council on Harmonization[26–28] (ICH) served as standards for manufacturing of biotechnology products.

In August 2002, the U.S. FDA announced a significant new initiative, Pharmaceutical Current Good Manufacturing Practices for the 21$^{st}$ Century, to enhance and modernize the regulation of pharmaceutical manufacturing and product quality. This initiative facilitates use of process analytical technologies,[29] Quality by design[30] and risk management[31] approaches to improve manufacturing. These principles can be applied to biotechnology products.[32] Platform approaches for manufacturing antibodies as well as other protein products may be of great benefit. Platforms may allow more rapid and efficient manufacturing of products. This advantage is even more important for manufacturing smaller amounts of a product, as would be the case during early clinical development or for orphan clinical indications. The greater use of disposables during manufacturing[33] may also allow for efficient delivery of smaller amounts of product. Contract and multiuse manufacturing facilities also increase options for the manufacturing of protein products.

Advances such as flexible, risk-based, multiscale manufacturing may facilitate the use of cytokines and other protein product in clinical trials. With greater ease in assuring high quality product, one potential barrier for the development of protein products can be lowered. Advances in manufacturing technology will be of great benefit in developing products for smaller populations and developing products that may only be used as part of a combination therapy.

## Inflammatory Cytokines

### Overview of Therapeutics Related to Inflammatory Cytokines

Advances in technology have reduced the likelihood of insurmountable manufacturing hurdles. However, manufacturing is only one of the hurdles in drug development. A product has to demonstrate appropriate safety and efficacy. Cytokines that can modulate inflammatory responses were discovered early, and their potential to treat autoimmune diseases, cancer, and infections led to great excitement. Many cytokines that can modulate inflammation and immunity have become marketed products. Some cytokines have large global sales, such as a number of IFN products. Interferon alpha products enhance the host response to some forms of hepatitis virus. Interferon beta products modulate inflammation in the autoimmune disease, multiple sclerosis. Other products block inflammatory cytokines and many of these have been very successful. Table 1 lists products related to inflammatory cytokines approved in the United States as of the writing of this review. Products shaded grey were included in the list of top 200 drugs in world-wide sales for 2006. The products in Table 1 generally have indications for autoimmunity, cancer or infectious disease. One exception is interleukin-11 (IL-11) which is used as a hematopoetic growth factor. However IL-11 also regulates inflammation and remodeling in the lungs[34] and has been studied in inflammatory diseases, such as Crohn's disease.[35] Thus it was included in this list. The pleiotropic nature[36] of cytokines can complicate the assignment of simple functional categories to cytokines.

Although Table 1 can be viewed as a great achievement, many stakeholders expected that a larger number of the known cytokines would become licensed products. In this sense,

Table 1 can be seen as only half-full, and it is important to consider the virtual rows that are empty. There are many inflammatory cytokines such as IL-10, and -12 that are not yet approved for clinical use despite great expectations. Cytokines that impact lymphocytes (lymphokines) have had a number of such candidates.

### Candidates in Waiting

In 1986, It was discovered that there are distinct populations of helper T-lymphocytes[37] and that these populations lead to very different biological outcomes. Type 1 helper T-lymphocytes secrete a set of cytokines, including IFN-$\gamma$, and were linked to delayed type hypersensitivity and cellular immune responses. Type 2 helper T-lymphocytes secrete a different set of cytokines, including IL-4 and -5, and were linked to immunoglobulin isotype switching. These distinct T-lymphocyte types are polarizing. A type 1 response leads to generation of more type 1 T-lymphocytes and vice versa. The nature of the response phenotype has a clear impact on the outcome of disease in a variety of models.[38]

IL-4 can be very important in the evolution of immune response phenotypes. It plays an important role in humoral immune responses, and was shown to have inhibitory effects on certain neoplasms.[39] IL-4 was studied in a number of indications, and although at least one study suggested a small benefit in non-small-cell lung cancer,[40] there have been no notable successes and IL-4 has not been approved for marketing.

The cytokine drivers of T cell polarization were further elucidated over time,[41] and IL-12 was found to be important in differentiating lymphocytes toward a type 1 cellular immune response. This suggested that IL-12 might be useful as a potential therapeutic for enhancing cellular immune responses against cancer or infections. An early study of IL-12 in patients with advanced renal cell carcinoma led to severe toxicity with 12 out of 17 patients hospitalized and two deaths.[9] This toxicity had

**TABLE 1.** Inflammatory Cytokine-related Protein Products Approved in the United States. These approvals are as of August 12, 2009. The grey shaded rows indicate the products that were included in the 2006 list of world's bestselling drugs from MedAdNews 200 – World's Best-Selling Medicines, MedAdNews, July 2007 (Wikipedia's wiki/List_of_top_selling_drugs). Products in italics were approved too late to be considered in 2006 global sales.

| US adopted name (USAN)[a] | Product type[b] | Cytokine or target class[c] | US approval date[d] | Disease targeted in initial indication[e] |
|---|---|---|---|---|
| Interferon-α-2b | Cytokine | IFN-α | 6/4/1986 | hairy cell leukemia |
| Interferon-α-2a | Cytokine | IFN-α | 6/4/1986 | hairy cell leukemia |
| Interferon-α-n3 | Cytokine | IFN-α | 10/10/1989 | genital warts |
| Aldesleukin | Cytokine | IL-2 | 5/5/1992 | metastatic renal cell carcinoma |
| Interferon-β-1b | Cytokine | IFN-β | 7/23/1993 | relapsing multiple sclerosis |
| Interferon-β-1a | Cytokine | IFN-β | 5/17/1996 | relapsing multiple sclerosis |
| Interferon alfacon-1 | Cytokine | IFN-α | 10/6/1997 | hepatitis C |
| Oprelvekin | Cytokine | IL-11 | 11/25/1997 | thrombocytopenia from chemotherapy for non-myeloid malignancies |
| Interferon-γ-1b | Cytokine | IFN-γ | 2/25/1999 | chronic granulomatous disease |
| Peginterferon-α-2b | Cytokine | IFN-α | 1/19/2001 | hepatitis C |
| Interferon-β-1a | Cytokine | IFN-β | 3/7/2002 | relapsing multiple sclerosis |
| Peginterferon-α-2a | Cytokine | IFN-α | 10/16/2006 | hepatitis C |
| Infliximab | Cytokine antagonist | TNF | 8/24/1998 | moderately to severely active or fistulizing Crohn's disease |
| Etanercept | Cytokine antagonist | TNF | 11/2/1998 | moderately to severely active rheumatoid arthritis |
| Anakinra | Cytokine antagonist | IL-1 | 11/14/2001 | moderately to severely active rheumatoid arthritis |
| Adalimumab | Cytokine antagonist | TNF | 12/31/2002 | moderately to severely active rheumatoid arthritis |
| *Rilonacept* | *Cytokine antagonist* | *IL-1* | *2/27/2008* | *cryopyrin-associated periodic syndromes (CAPS)* |
| *Certolizumab pegol* | *Cytokine antagonist* | *TNF* | *4/22/2008* | *Crohn's disease* |
| Denileukin diftitox | Cytokine fusion protein | IL-2 | 2/5/1999 | cutaneous T-cell lymphoma |
| Daclizumab | Cytokine receptor antagonist | IL-2R | 10/10/1997 | prevention of renal transplant rejection |
| Basiliximab | Cytokine receptor antagonist | IL-2R | 5/12/1998 | prevention of renal transplant rejection |

[a]Products are named using the established or nonproprietary name provided by the United States Adopted Name (USAN) Council. [b]Product types are classified as cytokines (native or modified cytokines that can modulate inflammatory responses), cytokine antagonists (protein products that can block an inflammatory cytokine such monoclonal antibodies and soluble cytokine receptors), cytokine fusion proteins (cytokines fused with another protein), and cytokine receptor Antagonists (protein products that block cytokine receptors from signaling, such as an antibody to a cytokine receptor). [c]The product class of cytokine or cytokine target, such as interleukins (IL), interferons (IFN) and tumor necrosis factor (TNF). [d]The approval date for the product in the United States based on the Drugs@FDA website. [e]The disease treated by the product in the initial label from the Drugs@FDA website. This information is intended for historical purposes. This column does not contain details on the indications or changes in the indications over time. The actual label should be consulted for current clinical indications and use.

not been seen in an earlier study where patients were exposed to a single test dose of IL-12 in advance of consecutive repeated dosing. This suggested that the toxicity could be avoided by more appropriate scheduling of doses. Models have supported this approach, and modest clinical studies have continued with IL-12. However these studies are primarily performed by academic investigators. Thus both support and drug supply can be limiting.

Returning to the theme of polarizing distinct sets of lymphocytes, the exact mechanisms of this polarization was a topic of great interest. One important mechanism was cross-regulation, with type 1 T-lymphocytes producing an inhibitor of type 2 differentiation and type 2 T-lymphocytes producing an inhibitor of type 1 differentiation. The IFN-γ produced by type 1 lymphocytes inhibits type 2 differentiation. However the inhibitor of type 1 differentiation produced by type 2 T-lymphocytes remained a mystery until it was found to be IL-10.[42] The ability of IL-10 to inhibit the cellular immune response associated with type 1 T-lymphocyte differentiation suggested a strong anti-inflammatory activity. Animal models supported this prediction in inflammatory bowel disease.[43,44] However, IL-10 treatment of Crohn's disease did not lead to significant clinical benefits despite decreasing a prominent marker of inflammation, that of nuclear factor kappa-light-chain-enhancer of activated B cells (NFκB) activation.[8]

The examples discussed above (IL-4, -12, and -10) are a number of cytokines that have not yet made it to approval. There are other candidates in similar situations, such as IL-6.[45] However, many cytokines that have not become approved products have provided excellent targets for development of very effective cytokine antagonists.

## Cytokine Antagonists

A soluble factor (lymphocyte activating factor) that potentiated lymphocyte activation was first described in 1972.[46] For decades factors such as endogenous pyrogen and leukocytic endogenous mediator were studied,[4] and in 1984 the genes for the corresponding proteins that accounted for many of these activities, IL-1 α and β, were cloned.[3] These molecules play an important role in inflammatory responses and pathology. Although IL-1 α did make it into clinical trials,[47] it has not yet been approved as

a therapeutic agent for any indication. However, the potent inflammatory effects of IL-1 suggested that antagonists of this cytokine could be of benefit in treating inflammatory conditions. This strategy led to development of two marketed products. An IL-1 receptor antagonist, anakinra, was approved as a secondary treatment of moderately to severely active rheumatoid arthritis in 2001.[5,48] In 2008 rilonacept, a fusion protein IL-1 antagonist, was approved for the rare cryopyrin-associated periodic syndromes (CAPS), including familial cold autoinflammatory syndrome (FCAS) and Muckle-Wells syndrome (MWS) in adults and children 12 and older.[6,49] The effectiveness of IL-1 antagonism in these orphan diseases suggests there may be rare diseases that could benefit from cytokine-related therapies.

Tumor necrosis factor (TNF; also referred to as TNF-α) was discovered in 1975, and as per its name, TNF can play an important role in tumor destruction.[50] A number of clinical studies have evaluated TNF as a single agent without much success.[51] Some studies using TNF in combination with other agents have led to toxicity. Strategies using TNF in combination therapies with regional perfusion may be of value for cancer indications. Tumor necrosis factor has also been recognized as an important component of chronic inflammatory diseases.[50] Antagonists of TNF have had great success in treating diseases such as rheumatoid arthritis and inflammatory bowel disease. Two of the three TNF antagonists that were marketed prior to 2006 were in the top 10 world's best-selling medicines in 2006, and all three were in the top 50.[52] These products have been great successes but can induce adverse events.[53,54]

## Approved Cytokine Products

Although some cytokines have led to therapies by being good drug targets, as indicated in Table 1, many cytokines are themselves approved products. In 1986 IFN-α was initially approved for treatment of hairy cell leukemia[55] and other oncology and antiviral indications followed. IFN-α moved from use as a single agent for chronic hepatitis C[56] to use in combination therapies. The addition of polyethylene glycol (PEG) to IFN allowed for improved pharmacokinetics.[57] PEG containing versions of IFN-α are in the top 200 world's best-selling medicines in 2006.[52] In 1993, the first IFN-β for use in treating multiple sclerosis was approved.[58] There are currently three approved IFN-β products for this indication, and all three were in the top 100 world's best-selling medicines in 2006.[52] In 1999 IFN-γ was approved for treatment of chronic granulomatous disease and later for osteopetrosis.[59] It is of note that, despite the important role of IFN-γ as a type 1 inflammatory cytokine, it has not yet been approved in treating infections in a broad population. Some cytokines may only provide benefit in specific syndromes where they or a downstream mediator are deficient. Even though IFN are filling important clinical needs and a number are being best-selling medicines, IFN did not meet the huge initial expectation as cancer therapeutics.[1,2]

IL-2 was the first T-lymphocyte growth factor discovered, and played a critical role in facilitating the in vitro growth of T-lymphocytes.[60] The ability of IL-2 to promote T-lymphocyte expansion made it an attractive protein for enhancing antitumor responses. IL-2 was approved in 1992 for treatment of metastatic renal cell carcinoma. IL-2 has also been studied in other tumors and different dosing regimens.[61,62] Although it is of benefit in a small subset of patients, IL-2 therapy elicits significant toxicities. In general, proinflammatory cytokines may have greater risks of toxicity than other cytokine products. Alternative therapeutic approaches, such as targeted delivery, may be of value. Strategies to enhance effectiveness have also been considered. These include use of IL-2 in combination therapies and a better understanding of the factors impacting efficacy.[63,64] In addition to the cytokine, IL-2, there are a number of related products. An IL-2 fusion protein has also been approved to

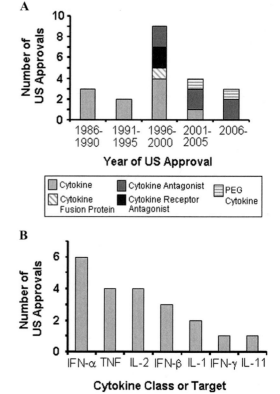

**Figure 1.** U.S. approvals of inflammatory cytokine-related protein products. These graphs are based on the data in Table 1. (**A**) U.S. approvals are broken down into five year intervals from 1986 onward. The product types in the key are defined as per Table 1, however a new product type is added. Cytokines linked to polyethylene glycol (PEG) are separated from unmodified cytokines. (**B**) U.S. approvals are broken down by the class of cytokine or targeted cytokine (ligand or receptor).

target diphtheria toxin to a T-lymphocyte lymphoma. Monoclonal antibodies targeting the IL-2 receptor have been approved to prevent renal allograft rejection (Table 1).

Inflammatory cytokine-related proteins have had a number of notable successes. However many cytokines with great expectations have not made it to market. In addition, a number of approved cytokine therapeutics have narrow indications despite striking biological effects in model systems. Figure 1A lists the approvals of cytokines in the United States over 5-year intervals. The number of native cytokines approved peaked between 1996 and 2000.

During that same time period, a cytokine fusion protein and a large number of cytokine antagonists were approved. From 2001 onward, cytokine antagonists make up the majority of approvals followed by modified cytokines (the addition of PEG to alter pharmacokinetics). This trend reflects a shift from the approval of cytokines as therapeutic agents to approval of cytokine antagonists as therapeutic agents. This may reflect greater challenges for the development of proinflammatory cytokines than for the development of inhibitors or modulators of inflammation.

Although Table 1 lists 21 products, all of these are related to seven cytokine classes. Table 1B shows that products related to IFN, TNF, and IL-2 account for more than 80% of approvals. As discussed earlier, many other classes of inflammatory cytokines have not yet led to approved therapeutics. The clinical experience with these cytokines suggests a number of approaches to leap beyond these hurdles.

## Future Approaches

### Predictive Science for Biology

One reason that expectations for many of these products were not met is that predictive science for biological systems is limited. Cytokines are not alone in this. Only a fraction of all drugs studied in the clinic make it to the market.[65] Despite this experience, every cytokine is anticipated as a potential "master regulator," able to control complex systems with ease. In the last 5 years, we are still calling cytokines such as IL-1, IL-10 and IFN-$\gamma$, master regulators.[66–68] However, it is of note that each of these claims is now in a particular context. Two decades ago, it was realized that cytokines are pleiotropic and redundant.[36] This suggests the biological impact of altering the level of a particular cytokine may be difficult to predict. It is likely that the more defined the context, the more accurate the prediction.

In addition to having overlapping functions, cytokines often share overlapping signaling pathways. A study looking at NF$\kappa$B signaling found a large number of genes that were positive and negative regulators.[69] The expression of these modulators was tissue specific. This again supports the importance of context. The results of this study were integrated into maps of signal transduction. In Figure 2A, a portion of the signaling map of TNF is shown. A small section of this map containing two positive regulators was expanded to show the complex nature of the interactions in this pathway (Fig. 2B). Since NF$\kappa$B plays an important role in many other cytokine signaling cascades, the nodes in this complex circuitry overlap with other signaling pathways. Combining this with tissue and state-specific variations makes for a very complex system. Use of genomic and proteomic approaches is critical in developing better approaches to understand and predict cytokine behavior in complex biological systems.

## Newer Cytokines

Many cytokines are still offering promises of therapeutic advances, such as IL-7 and -15.[70] The discovery of IL-17 and the T-lymphocyte subset that produces it has led to renewed enthusiasm for the use of cytokines as therapeutic agents. This helper T-lymphocyte subset is a relatively new addition to the type 1 and type 2 subsets defined in 1986 and may represent an important cell type in clearing pathogens and inducing autoimmune tissue inflammation.[71] Other cytokines, such as IL-21 and -23 are linked to this T-lymphocyte subset.[72,73] Novel IFN such as IFN-$\lambda$ (IL-29) that target tissues more selectively than IFN-$\alpha$ may also offer therapeutic advantages.[74] TNF is part of a large family of cytokines that can also lead to new therapies.[75]

## Picking Winning Indications

Assuming toxicities are not limiting, some of these cytokines may have such broad

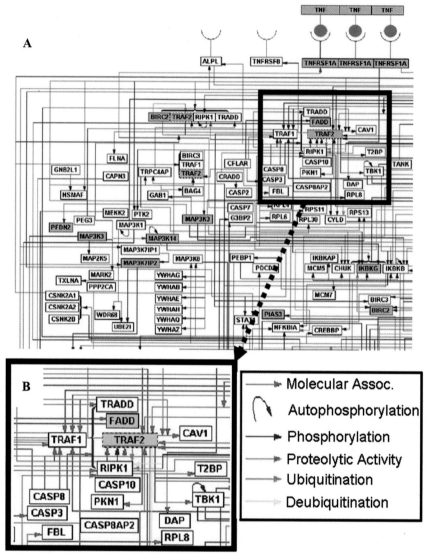

**Figure 2.** Cytokine signaling networks are complex and interlinked. (**A**) is a section of a map for TNF signaling. Shaded boxes represent genes identified as modulators of NFκB, an important signaling pathway for a variety of cytokines, antigen receptors, growth factors, and pathogen associated molecules. (**B**) is a small section of the pathway map is expanded. Identified positive regulators of NFkB are shown in green boxes. Interactions between gene products are indicated by arrows and identified at the right. This figure was adapted from Ref. 69, Supplemental Figure 5, with permission.

biological impact that they will work in a variety of large market indications. However, if we learn from history, it is unlikely they will all be successful. It would be useful to take advantage of available knowledge in systems biology for choosing indications to study and for designing appropriate therapeutic regimens. In addition, capturing and evaluating as much information as possible during preclinical and clinical development may provide additional help in choosing indications and designing novel regimens. Making the right choices for clinical development is often the difference between success and failure. In the case of an antibody targeting vascular endothelial growth factor (VEGF), a phase 3 trial for patients with previously treated breast cancer did not demonstrate an improvement in progression-free or overall survival.[76] However when used in patients that had not been previously treated, progression-free survival improved.[77] In addition, overall survival improved for colorectal and lung cancer patients. In this case, a difference in study design had a profound effect on the clinical outcome.

The right indication may not be the indication with the largest market. The examples of IFN-γ and the IL-1 antagonist, rilonacept, suggest products that have broad biological activity may be the most appropriate therapy for rare or orphan diseases. Cytokine-related products that may not find a large market indication may have tremendous benefits for the right population of patients. As indicated above, systems biology may be a very helpful tool in finding the right indication for a product. The historical industry approach to drug development has been to look for blockbuster drugs. It would be useful for drug development portfolios to include more products that target smaller indications. Development of drugs for defined populations in conjunction with systems biology approaches may decrease attrition. A higher success rate might also compensate for smaller markets. In addition, when a drug fails due to lack of efficacy or due to an indication-specific toxicity, it would be useful to consider other indications. Understanding clinical failures may be very useful in devising a winning development plan.

## Strategies for Development

A cytokine product may not find an indication despite the use of systems biology and an exhaustive search among rarer conditions. That does not mean it cannot be of therapeutic value. If a protein has potent biological activity, there may be alternative approaches to

capitalize on. One is to generate an antagonist of that cytokine or its cognate receptor protein. That strategy has clearly been successful for cytokines such as TNF and IL-1. Antagonists can be soluble cytokine receptors, altered ligands, fusion proteins, or monoclonal antibodies. Cytokine antagonists continue to be developed. In Europe, an antibody targeting the p40 subunit shared by IL-12 and -23 was approved.[78] Cytokine receptors can also be targeted, such as antibodies to the IL-2 receptor alpha chain. An antibody to the IL-6 receptor was also recently approved in Europe.[79]

Another approach is to modify the cytokine to allow for successful clinical development. In the case of INF-α, the addition of a synthetic carbohydrate, polyethylene glycol (PEG), greatly improved product pharmacokinetics. Although in this case, the modification improved already marketed products, there may be cases where a particular pharmacokinetic profile would be necessary for initial approval. Fusion proteins[80] can also be used for altering pharmacokinetics or adding function. Proteins can be truncated to remove an undesirable functional activity. Glycosylation can be altered to impact protein trafficking.[81] Directed mutagenenesis can been used to impact binding, effector functions, glycosylation, and pharmacokinetics.[82] Evolutionary approaches can be used on specific areas of a sequence[83] or to shuffle large segments of a protein.[84] However, these modifications may impact the immunogenicity of a protein. Immunogenicity is a concern for all protein products and can affect safety and efficacy. Changes in structure and especially changes that lead to novel primary sequence can alter immunogenicity. Increased immunogenicity may diminish efficacy or result in adverse events that can hinder drug development.

Even if, despite consideration of modifications, a biologically active cytokine has no likely indication, it may be of therapeutic value in combination with another agent. Considering the complexity and overlap of signaling pathways, it is possible that a product may be ineffective due to redundancy and or homeostatic mechanisms. Understanding the cytokine signaling pathways may illustrate situations where combination therapies will defeat redundancy or homeostatic mechanisms. However, it is critical to be aware that such combinations may increase the risk of adverse events through an exaggerated mode of action. Combinations of cytokines with different toxicities may be used to increase safety by achieving a therapeutic endpoint with lower doses of each cytokine. Finally, the appropriate delivery of a cytokine to the correct tissue or organ may make the difference between clinical success and failure. Pro-inflammatory cytokines, such as IL-2, IL-12, and TNF, could be more effective and less toxic if administered in a more tissue-selective manner.

Decisions regarding drug development of cytokine products should not just be based on the general biology of the molecule but on the biology of specific disease states. Figure 3A considers diseases based on three dimensions. The Y-axis covers the number of causal entities or factors that are necessary for disease expression. These may be genes, epigenetic factors or environmental exposures. This data would inform the basic choice of agonist versus antagonist. It would also provide information on whether a single agent approach is likely to work or a combination therapy is needed. A deficiency disease of a single protein would have a low score on this axis. The X-axis covers the localization of these disease factors. Disease factors may operate systemically or may occur in a particular tissue, cell type or cell compartment. This axis would provide information on whether a targeted delivery system is needed. This could be a structural modification of glycosylation to target a lysosomal compartment[81] or an implantable device to deliver a therapeutic product to a primary tumor. A disease causing factor that is always accessible to a systemic therapy would have a low score on this axis. The Z axis covers the levels and kinetic profiles of the factors necessary to produce disease. For type 1 diabetes, exposures to

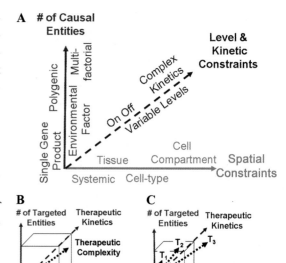

**Figure 3.** Complex diseases may require complex therapeutics. (**A**) Three axes of disease complexity. The Y-axis covers the number of causal entities or factors (necessary for disease phenotype). The X-axis covers the localization of these factors, and the Z-axis covers the levels and kinetics of these factors. (**B**) A therapeutic perspective of disease complexity (Y-number of targets, X-location of targets, Z-quantitative and kinetic considerations). The necessary therapeutic complexity can be mapped in this three dimensional space. Diseases evolve over time and that is important to consider. (**C**) Different therapeutic complexities at three different time points (T1, T2, and T3) in an evolving disease.

carbohydrates and other metabolic stresses are variable. Thus, even if this disease is viewed as a primary insulin deficiency, the replacement of insulin requires very complex dosing and kinetics. Delivery systems[85] that can correctly account for this can improve treatment of this disease. Figure 3B illustrates that disease complexity directly impacts therapeutic complexity. As described above, the causal factors, their localization and kinetics will impact the therapeutic complexity needed to treat the disease. As diseases evolve over time, the complexity of effective therapeutics may also need to evolve (Fig. 3C). An early stage of a chronic inflammatory disease may only require a single agent. A later stage may require additional agents.

## Conclusions and Future Prospects for Cytokines

Inflammatory cytokine-related products have had great successes in treating diseases such as rheumatoid arthritis, multiple sclerosis, hepatitis, Crohn's disease, and some malignancies. However, many cytokines with powerful biological activities have not yet succeeded as therapeutics. There are always barriers in drug development, and with appropriate efforts and time, they can often be overcome. At the beginning of cytokine therapeutics, manufacturing of sufficient quantities of high-quality product was a huge hurdle. But we have leaped beyond that hurdle. Manufacturing is still a challenge, but there are still many barriers. Predicting clinical success and choosing the right indication(s) and therapeutic regimens for drug development are the current challenges. Systems biology and better knowledge management can help. These can also suggest protein modifications to increase the likelihood of clinical success. Improved strategies for drug delivery, from glycoengineering to nanoparticles, can increase the odds of success. All of these approaches have been started and some have led to approved products. They were all discussed at the NYAS/FDA meeting on March 26–27, 2009.[10] Although there was great interest in these approaches, a number of practical barriers were also discussed. These included limited accessibility to cytokine products to explore new indications and combination treatments. The lack of standards for cytokine measurements and systems biology data sets was felt to be problematic. Communication between industry, academia and regulatory agencies could be improved. The culture of drug development was also viewed as a problem. The focus on blockbuster drugs and the lack of second chances for failed products may not be the best long-term strategy for drug development. In closing, cytokine therapeutics have leaped over some hurdles though more lie ahead. There is much to be learned from past experiences with many of these products. By taking the right measured steps over time, we will be in a better position to turn barriers and hurdles into leaps and bounds.

## Disclaimer

The opinions represent those of the authors. The mentioned products and examples are chosen to illustrate product development. In some cases targeted diseases or other product attributes are noted. No endorsement or judgment on clinical use beyond the product labeling on U.S. approved products should be inferred.

## Acknowledgments

Thanks to Hadas Kozlowski for valuable editing suggestions. Since the drafting of this article, additional cytokine-based therapeutics were approved by the U. S. FDA. These products are another IFN-β-1b and ustekinumab, a monocolonal antibody targeting the p40 subunit of the IL-12 and IL-23 cytokines.

## Conflicts of interest

The authors declare no conflicts of interest.

## References

1. Sikora, K. 1980. Does interferon cure cancer? *Br. Med. J.* **281:** 855–858.
2. Kirkwood, J.M. & M.S. Ernstoff. 1984. Interferons in the treatment of human cancer. *J. Clin. Oncol.* **2:** 336–352.
3. Dinarello, C.A. 1994. The IL–1 family: 10 years of discovery. *FASEB J.* **8:** 1314–1325.
4. Dinarello, C.A. 1984. IL–1. *Rev. Infect. Dis.* **6:** 51–95.
5. 2002. Anakinra (Kineret) for rheumatoid arthritis. *Med. Lett. Drugs Ther.* **44:** 18–19.
6. Kapur, S. & M.E. Bonk. 2009. Rilonacept (Arcalyst), an IL–1 trap for the treatment of cryopyrin-associated periodic syndromes. *P. T.* **34:** 138–141.
7. Foster, D. *et al.* 2004. Cytokine-receptor pairing: accelerating discovery of cytokine function. *Nat. Rev. Drug Discov.* **3:** 160–170.

8. Bickston, S.J. & F. Cominelli. 2000. Recombinant IL-10 for the treatment of active Crohn's disease: lessons in biologic therapy. *Gastroenterology* **119:** 1781–1783.

9. Leonard, J.P. *et al.* 1997. Effects of single-dose IL–12 exposure on IL–12-associated toxicity and interferon-gamma production. *Blood* **90:** 2541–2548.

10. NYAS. 2009. Cytokine Therapies: Novel Approaches for Clinical Indications. eBriefing.

11. Cantell, K. *et al.* 1974. Human leukocyte interferon: production, purification, stability, and animal experiments. *In Vitro Monogr.* 35–38.

12. Gibbs, C.J., Jr. *et al.* 1985. Clinical and pathological features and laboratory confirmation of Creutzfeldt-Jakob disease in a recipient of pituitary-derived human growth hormone. *N. Engl. J. Med.* **313:** 734–738.

13. Singer, M. & P. Berg. 1976. Recombinant DNA: NIH guidelines. *Science* **193:** 186–188.

14. Peters, M. *et al.* 1986. Immunologic effects of interferon-alpha in man: treatment with human recombinant interferon-alpha suppresses in vitro immunoglobulin production in patients with chronic type B hepatitis. *J. Immunol.* **137:** 3147–3152.

15. Smith, S.L. 1996. Ten years of Orthoclone OKT3 (muromonab-CD3): a review. *J. Transpl. Coord.* **6:** 109–119.

16. Spellman, M.W. *et al.* 1989. Carbohydrate structures of human tissue plasminogen activator expressed in Chinese hamster ovary cells. *J. Biol. Chem.* **264:** 14100–14111.

17. Cox, M.M. 2008. Progress on baculovirus-derived influenza vaccines. *Curr. Opin. Mol. Ther.* **10:** 56–61.

18. Kling, J. 2009. First US approval for a transgenic animal drug. *Nat. Biotechnol.* **27:** 302–304.

19. Yano, A. & M. Takekoshi. 2004. Transgenic plant-derived pharmaceuticals – the practical approach? *Expert. Opin. Biol. Ther.* **4:** 1565–1568.

20. O'Connor, J.V. 1993. The use of peptide mapping for the detection of heterogeneity in recombinant DNA-derived proteins. *Biologicals* **21:** 111–117.

21. Cho, A. & D. Normile. 2002. Nobel Prize in Chemistry. Mastering macromolecules. *Science* **298:** 527–528.

22. ICH. 2004. Q5E Comparability of Biotechnological/Biological Products Subject to Changes in their Manufacturing Process.

23. U.S. FDA. 1996. FDA Guidance Concerning Demonstration of Comparability of Human Biological Products, Including Therapeutic Biotechnology-derived Products.

24. U.S. FDA. 1997. Guidance for Industry: Changes to an Approved Application for Specified Biotechnology and Specified Synthetic Biological Products.

25. U.S. FDA. 1997. Points to Consider in the Manufacture and Testing of Monoclonal Antibody Products for Human Use.

26. ICH. 1999. Q6B Specifications: Test Procedures and Acceptance Criteria for Biotechnological/Biological Products.

27. ICH. 1995. Q5B Analysis of the Expression Construct in Cells Used for the Production of r-DNA Derived Protein Products and Q5C Stability Testing for Biotechnology/Biological Products.

28. ICH. 1997. Q5A Viral Safety Evaluation of Biotechnology products derived from lines of Human or Animal Origin and Derivation and Q5D Characterization of Cell Substrates Used for the Production of Biotechnology/Biological Products.

29. U.S. FDA. 2004. Guidance for Industry PAT—A Framework for Innovative Pharmaceutical Development, Manufacturing, and Quality Assurance.

30. ICH. 2008. Q8(R1) Pharmaceutical Development Revision 1.

31. ICH. 2006. Q9: Quality Risk Management.

32. Rathore, A.S. & H. Winkle. 2009. Quality by design for biopharmaceuticals. *Nat. Biotechnol.* **27:** 26–34.

33. Rao, G., A. Moreira & K. Brorson. 2009. Disposable bioprocessing: the future has arrived. *Biotechnol. Bioeng.* **102:** 348–356.

34. Zheng, T. *et al.* 2001. IL-11: insights in asthma from overexpression transgenic modeling. *J. Allergy Clin. Immunol.* **108:** 489–496.

35. Herrlinger, K.R. *et al.* 2006. Randomized, double blind controlled trial of subcutaneous recombinant human IL–11 versus prednisolone in active Crohn's disease. *Am. J. Gastroenterol.* **101:** 793–797.

36. Paul, W.E. 1989. Pleiotropy and redundancy: T cell-derived lymphokines in the immune response. *Cell* **57:** 521–524.

37. Mosmann, T.R. *et al.* 1986. Two types of murine helper T cell clone. I. Definition according to profiles of lymphokine activities and secreted proteins. *J. Immunol.* **136:** 2348–2357.

38. Mosmann, T.R. & S. Sad. 1996. The expanding universe of T-cell subsets: Th1, Th2 and more. *Immunol. Today* **17:** 138–146.

39. Maher, D.W. *et al.* 1991. Human IL–4: an immunomodulator with potential therapeutic applications. *Prog. Growth Factor Res.* **3:** 43–56.

40. Vokes, E.E. *et al.* 1998. A phase II study of recombinant human IL–4 for advanced or recurrent non-small cell lung cancer. *Cancer J. Sci. Am.* **4:** 46–51.

41. Seder, R.A. & W.E. Paul. 1994. Acquisition of lymphokine-producing phenotype by CD4+ T cells. *Annu. Rev. Immunol.* **12:** 635–673.

42. Bashyam, H. 2007. Th1/Th2 cross-regulation and the discovery of IL-10. *J. Exp. Med.* **204:** 237.

43. Grool, T.A. *et al.* 1998. Anti-inflammatory effect of IL–10 in rabbit immune complex-induced colitis. *Scand. J. Gastroenterol.* **33:** 754–758.

44. Kennedy, R.J. *et al*. 2000. IL- 10-deficient colitis: new similarities to human inflammatory bowel disease. *Br. J. Surg.* **87:** 1346–1351.

45. Stouthard, J.M. *et al*. 1996. Recombinant human IL-6 in metastatic renal cell cancer: a phase II trial. *Br. J. Cancer* **73:** 789–793.

46. Gery, I., R.K. Gershon & B.H. Waksman. 1972. Potentiation of the T-lymphocyte response to mitogens. I. The responding cell. *J. Exp. Med.* **136:** 128–142.

47. Smith, J.W., 2nd *et al*. 1992. The toxic and hematologic effects of IL–1 alpha administered in a phase I trial to patients with advanced malignancies. *J. Clin. Oncol.* **10:** 1141–1152.

48. Hallegua, D.S. & M.H. Weisman. 2002. Potential therapeutic uses of IL- 1 receptor antagonists in human diseases. *Ann. Rheum. Dis.* **61:** 960–967.

49. Hoffman, H.M. 2009. Rilonacept for the treatment of cryopyrin-associated periodic syndromes (CAPS). *Expert Opin. Biol. Ther.* **9:** 519–531.

50. Clark, I.A. 2007. How TNF was recognized as a key mechanism of disease. *Cytokine Growth Factor Rev.* **18:** 335–343.

51. Brophy, L. 2001. Tumor necrosis. In Biotherapy A Comprehensive Overview. P.T. Reiger, ed. Jones and Bartlett. Sudbury, MA.

52. 2007. 200 – World's Best-Selling Medicines. MedAd-News, July.

53. Leombruno, J.P., T.R. Einarson & E.C. Keystone. 2009. The safety of anti-tumour necrosis factor treatments in rheumatoid arthritis: meta and exposure-adjusted pooled analyses of serious adverse events. *Ann. Rheum. Dis.* **68:** 1136–1145.

54. Huggett, B. 2008. FDA probes TNF blockers. *Nat. Biotechnol.* **26:** 845.

55. 1986. Interferon for treatment of hairy-cell leukemia. *Med. Lett. Drugs Ther.* **28:** 78–79.

56. Baron, S. *et al*. 1991. The interferons. Mechanisms of action and clinical applications. *JAMA* **266:** 1375–1383.

57. Saadeh, S. & G.L. Davis. 2004. The evolving treatment of chronic hepatitis C: where we stand a decade out. *Cleve. Clin. J. Med.* **71**(Suppl 3): S3–7.

58. 1993. Interferon beta-1B for multiple sclerosis. *Med. Lett. Drugs Ther.* **35:** 61–62.

59. Shankar, L., E.A. Gerritsen & L.L. Key Jr. 1997. (No title available.) *BioDrugs* **7:** 23–29.

60. Gillis, S. 1983. IL- 2: biology and biochemistry. *J. Clin. Immunol.* **3:** 1–13.

61. Atkins, M.B. *et al*. 1999. High-dose recombinant IL- 2 therapy for patients with metastatic melanoma: analysis of 270 patients treated between 1985 and 1993. *J. Clin. Oncol.* **17:** 2105–2116.

62. Gale, D.M. & P. Sorokin. 2001. Hurdles and leaps for protein therapeutics: Cytokines and inflammation. In Biotherapy A Comprehensive Overview. P.T. Reiger, ed.: 1–15. Jones and Bartlett. Sudbury, MA.

63. Romo de Vivar Chavez, A. *et al*. 2009. The biology of IL–2 efficacy in the treatment of patients with renal cell carcinoma. *Med. Oncol.* **26**(Suppl 1): 3–12.

64. Smith, K.A. 1997. Rational IL–2 therapy. *Cancer J Sci Am.* **3**(Suppl 1): S137–S140.

65. DiMasi, J.A., R.W. Hansen & H.G. Grabowski. 2003. The price of innovation: new estimates of drug development costs. *J. Health Econ.* **22:** 151–185.

66. Basu, A., J.K. Krady & S.W. Levison. 2004. IL–1: a master regulator of neuroinflammation. *J. Neurosci. Res.* **78:** 151–156.

67. Couper, K.N., D.G. Blount & E.M. Riley. 2008. IL-10: the master regulator of immunity to infection. *J. Immunol.* **180:** 5771–5777.

68. McLaren, J.E. & D.P. Ramji. 2009. Interferon gamma: a master regulator of atherosclerosis. *Cytokine Growth Factor Rev.* **20:** 125–135.

69. Halsey, T.A. *et al*. 2007. A functional map of NFkappaB signaling identifies novel modulators and multiple system controls. *Genome Biol.* **8:** R104.

70. Kim, H.R. *et al*. 2008. IL-7 and IL-15: biology and roles in T-Cell immunity in health and disease. *Crit. Rev. Immunol.* **28:** 325–339.

71. Korn, T. *et al*. 2009. IL-17 and Th17 cells. *Annu. Rev. Immunol.* **27:** 485–517.

72. Neurath, M.F. 2007. IL-23: a master regulator in Crohn disease. *Nat. Med.* **13:** 26–28.

73. Deenick, E.K. & S.G. Tangye. 2007. Autoimmunity: IL-21: a new player in Th17-cell differentiation. *Immunol. Cell Biol.* **85:** 503–505.

74. Kotenko, S.V. *et al*. 2003. IFN-lambdas mediate antiviral protection through a distinct class II cytokine receptor complex. *Nat. Immunol.* **4:** 69–77.

75. Aggarwal, B.B. *et al*. 2002. The role of TNF and its family members in inflammation and cancer: lessons from gene deletion. *Curr. Drug Targets Inflamm Allergy* **1:** 327–341.

76. Genentech. 2003. A Study of rhuMAb VEGF (Bevacizumab) in Combination with Chemotherapy in Patients with Previously Treated Breast Cancer: rhuMAb VEGF-CSR AVF2119g Final. Clinical Study Report Synopsis.

77. Jain, R.K. *et al*. 2006. Lessons from phase III clinical trials on anti-VEGF therapy for cancer. *Nat. Clin. Pract. Oncol.* **3:** 24–40.

78. Reich, K., U. Yasothan & P. Kirkpatrick. 2009. Ustekinumab. *Nat. Rev. Drug Discov.* **8:** 355–356.

79. Scheinecker, C. *et al*. 2009. Tocilizumab. *Nat. Rev. Drug Discov.* **8:** 273–274.

80. 1998. Etanercept marketed for moderate, severe rheumatoid arthritis. *Am. J. Health Syst. Pharm.* **55:** 2593.

81. Zhu, Y. *et al.* 2004. Conjugation of mannose 6-phosphate-containing oligosaccharides to acid alpha-glucosidase improves the clearance of glycogen in pompe mice. *J. Biol. Chem.* **279:** 50336–50341.

82. Elliott, S. *et al.* 2003. Enhancement of therapeutic protein in vivo activities through glycoengineering. *Nat. Biotechnol.* **21:** 414–421.

83. Wu, H. *et al.* 2007. Development of motavizumab, an ultra-potent antibody for the prevention of respiratory syncytial virus infection in the upper and lower respiratory tract. *J. Mol. Biol.* **368:** 652–665.

84. Chang, C.C. *et al.* 1999. Evolution of a cytokine using DNA family shuffling. *Nat. Biotechnol.* **17:** 793–797.

85. Brunton, S. 2008. Insulin delivery systems: reducing barriers to insulin therapy and advancing diabetes mellitus treatment. *Am. J. Med.* **121:** S35–S41.